Laboratory Manual
for

ELECTRICITY FOR HVACR

Nicholas Griewahn
Northern Michigan University

Samuel Shane Todd
Ogeechee Technical College

PEARSON

Boston   Columbus   Indianapolis   New York   San Francisco   Upper Saddle River

Amsterdam   Cape Town   Dubai   London   Madrid   Milan   Munich   Paris   Montreal   Toronto

Delhi   Mexico City   Sao Paulo   Sydney   Hong Kong   Seoul   Singapore   Taipei   Tokyo

Editorial Director: Vernon R. Anthony
Editorial Assistant: Nancy Kesterson
Director of Marketing: David Gesell
Senior Marketing Coordinator: Alicia Wozniak
Marketing Assistant: Les Roberts
Program Manager: Maren L. Miller
Operations Specialist: Deidra Skahill

Senior Art Director: Diane Y. Ernsberger
Development Editor: Leslie Lahr
Textbook Cover Designer: Suzanne Behnke
Manual Cover Designer: Integra
Cover Art: Fieldpiece Instruments
Printer/Binder: LSC Communications
Cover Printer: LSC Communications

Credits and acknowledgments borrowed from other sources and reproduced, with permission, in this textbook appear on the appropriate page within text.

PEARSON

ISBN 10: 0-13-512536-7
ISBN 13: 978-0-13-512536-6

4 2023

A Note to Students

We are happy you have chosen to expand your knowledge of heating, air conditioning, and refrigeration by using this lab manual. It is designed to complement the textbook *Electricity for HVACR*, first edition, and the unit numbers of the lab manual coincide with the text. This lab manual is not designed to give you all the practical experience you will need, but it is a start and everyone in the HVACR field must start somewhere. Although electrical components have evolved, the fundamental of electricity has not changed. Electrical meters have changed a great deal, but what these meters check has not: voltage, amperage, and ohms are the same now as they have always been. The labs in this manual are an accumulation of knowledge from several people with many years of experience in HVACR. It is important for you to know that these people are still learning the trade. The HVACR field offers an opportunity for lifelong learning and professional growth.

A note on using this manual: Many students want to skip the reading and get to the hands-on activities, but these activities are useless without background knowledge or theory. You need to take your time and read all of the content of the labs.

It is our hope that you gain much knowledge from this lab manual and that you enjoy learning about HVACR.

Best of luck in your HVACR career,

Nick Griewahn

Shane Todd

PREFACE

Organization

Each unit of this manual includes summary material from your text, key terms associated with the unit, and most have one or more labs (included for selected appropriate units). The format of each lab in this manual is consistent and includes the following features:

Laboratory Objective, which provides the overarching goal of the lab.

Text Reference, which directs students and instructors to the text material most closely associated with the lab.

Required Materials, which enable both student and instructor to prepare for lab sessions and use lab time productively.

Safety Requirements, which offer precautions and safety practices that students should follow in the lab and in the field.

Introduction, which provides students with valuable background information pertaining to the lab based on the authors' years of experience in the field and in the classroom.

Procedure, which carefully walks students through clearly described steps required to complete the lab in a satisfactory manner; steps often require recorded results and observations.

Questions, which test student understanding of the key ideas and concepts within the lab and the associated text unit.

Safety Considerations and Equipment

Many of the assignments found within this manual involve students working around and directly with electricity of several voltages. You will want to follow general safety precautions as well as the more specific safety guidelines that are found with each assignment. Electricity can be dangerous, but some common sense precautions and thinking before you act when working around electrical components will help ensure your safety. With this in mind, many people have been working around electricity for many years with no problems. Electricity is something to be respected, not feared.

The tools you will use with electrical and HVACR can be hazardous if not handled and used properly. Hand tools should only be used for their intended purpose. Vise grips are intended to handle hot parts or hold two objects together, they are not wrenches or hammers. The same way that pencils, computers, and calculators are tools of the trade for a mathematician, screwdrivers, wrenches, and wire strippers are the tools of your trade. You would expect a professional of any type to have top quality tools. You should try not to save a few dollars by purchasing cheap value brand tools. These tools are what you will be using to make your living and they should last for many years. Inexpensive tools are not only frustrating, but they can be dangerous as well. In general, the more money you spend, the better quality you will receive. In particular, high quality digital multimeters or volt-ohm meters are of utmost importance. In many cases measurements from this meter will tell you if a circuit is safe to touch and it can help you accurately diagnose very expensive and critical systems that the customer cannot afford to replace should you damage it.

There are some instances where you do not have a choice but to work around live circuits and work very closely to them. Voltage and current measurements you will take can only be taken when power is applied and you must learn to work safely around energized circuits. When the situation demands that you work with live electricity you should have a regular plan to go about your work. First, never work on a live circuit that you do not have a means of turning off quickly in the event of an emergency. Always know where the nearest disconnect or power switch is located and be prepared to turn it off quickly. Work with proper safety gear for the voltage you are working with. Low voltage circuits with 24 VAC or less are the least hazardous and you won't need anything more than gloves and safety glasses to work with them. Higher line voltages that are above 24 VAC demand caution and you should use safety glasses and gloves at all times when working with them. With voltages of 460 VAC and higher, try to work with one hand away from the circuit. Should you get a shock with both hands on the wiring the current could travel through your chest instead of only through your fingers and arm. Some facilities that operate with 460 VAC and higher require the use of high calorie suits and arc flash protection equipment. Arc flash is when voltage jumps from the source to a ground. The flash that is generated is hotter than the surface of the sun and can badly burn someone. These flashes can be unpredictable and can occur during power surges or equipment problems. When using power tools always use double insulated tools or tools connected properly to a ground conductor. The ground is designed to direct current to the path of least resistance, which is usually around your body instead of through it.

Equipment Notes for Instructors

There is certain equipment that will be required for these labs that are used several times. Every effort was made in this lab manual to accommodate all HVACR laboratory classrooms and to use the equipment they currently possess. This section is intended to address some specifics of this equipment and some tips to help instructors get the most out of the equipment that will be needed.

Basic Wiring Trainers: Used for labs 2.1, 2.2, 3.2, 9.2, 9.4, 26.1

This is a basic trainer that can be used for many electrical training purposes. The trainer should contain at least four SPST toggle switches, two SPDT toggle switches, three 60 Watt light bulbs, three 100 Watt light bulbs screwed into cleat type receptacles, a fused 115 VAC power source, and a disconnect switch. The substrate for the trainer can be something as inexpensive and a 24" x 24" sheet of plywood with the components mounted to it in an orderly fashion. You will also need an assortment of wires for students to connect components together for their projects. Banana plugs work best, but insulated alligator clips can also work well and can be purchased at relatively low cost from an electronics supplier. Relays will also be required for several labs and this trainer provides the circuitry needed to wire them easily. The relays should be MARS 64 type or similar with 24 volt coils. You should have at least one two-pole contactor, two SPST relays, and two SPDT relays. For the 24-volt coils you will also need a 40 VAC transformer. The relays and transformer can be mounted on the same board as the light bulbs and toggle switches, or they can be a separate accessory board that is only used for the few labs that need them. It is recommended that students be charged with building the boards they will use for the academic term and then disassembling them at the end. By building the trainers themselves students will be more familiar with the components and where they are located. Also, since they will be using them several times through the entire semester, building it themselves leaves them a bit of responsibility since they will have to deal with any flaws down the road.

HVACR system trainers: Used for most labs

All well-equipped labs should have a good variety of systems to work with. Most manufacturers are willing to work with schools to provide them with modern systems at low or no cost to the school since the manufacturer benefits by impressing their brand on tomorrow's technicians. There are situations, such as in Labs 6.1 or 27.1, where the system will need some manuals and specifications from the manufacturer. Nameplate data including model and serial numbers, and basic electrical data is a must. In many cases in the field these documents cannot be located. Encourage your students to use electronic resources that they would be able to access in the field. Internet, supply houses, and manufacturer technician hotlines are all good resources to find information they need. The specific systems required for using this manual to its full extent are split system air conditioners, packaged air conditioners, electric heating systems, air source heat pumps, and gas-fired forced air furnaces. Other basic refrigeration systems and hydronic heating and cooling systems are also useful, but not required.

The systems should be in working order unless specified in the lab activity. Labs 7.1, 9.4, 14.1, 14.2, 16.1, 25.1 and 27.1 have special instructions to set up before the students use them. Inserting problems, including faulty components or improper settings, is essential in providing students with high-quality learning opportunities. These "bugs" should be something similar to what they would encounter in the field, and while it might take more time and energy to set up, a real system with a bug trumps any simulation they could use. These real-life systems are incorporated into lab activities as often as they practically could be.

Here are some lab specific requirements and some suggestions on how to set up the equipment and components.

Lab 9.1 – Some of the relays, contactors, and motor starters that are being checked by the students should have open coils for them to find.

Lab 9.4 – Some suggestions for control circuit bugs:

- Faulty high or low pressure switches

- Faulty discharge temperature limits

- Faulty time delay controls

- Open transformer primary or secondary coils

- Open contactor coil

- Try to avoid low or high pressure problems (with the pressure safeties being good)

Lab 14.1 – For the motors installed in the systems try to select a variety of motors, not just fan motors. Pump motors and compressor motors should be included where possible.

Lab 14.2 – For the capacitor start motor required, the motor should <u>not</u> be using a contactor or motor starter. All three types of motors should also contain some accessory safety controls like thermal overloads or pressure safeties.

Lab 14.3 – The watt meter requirement is more of a suggestion. A watt meter helps students understand the difference between power and current and how each is affected by the load on a motor.

Lab 16.1 – The ECM and PSC motor in the two systems used should be approximately the same size in horsepower and driving a similarly sized blower wheel.

Lab 25.1 – For the heating and cooling system with a fault make the fault a simple one like a blown fuse, open switch, or obviously broken connection so that <u>one</u> load does not function.

Lab 26.1 – Some suggestions for these faulty components: relays with open coils, bad contacts, faulty light bulbs, faulty switches, broken wires, etc.

Lab 27.1 – All of the systems, operational or not, should have adequate literature that accompanies them including installation guidelines, electrical data, and diagrams. If the unit does not have this material, make sure the students have the resources to find the information they need.

To locate a supply of bugs for these systems, contact local contractors who are usually more than willing to save a few bad parts to help train future technicians. Other bugs, like open coils, can be created fairly easily by clipping the coil wires, just make sure you conceal it as much as possible. Creating a library of faulty components for students to find can be challenging, but fun.

Tools of the Trade

As mentioned, the tools you select for work should be quality tools that you can work well with. There are certain tools that need some additional explanation, however. Most schools require that you provide your own multimeter. The brands and options multimeters have are quite dizzying. Getting a meter from a reputable HVACR supplier or distributor is a good idea. These suppliers usually only carry DMMs that are designed for use in HVACR and they will help you with questions and any warranty issues down the road. Buying a multimeter from an auto parts store will likely result in getting a meter more geared to DC power and automotive use.

As far as DMM options are concerned here are a few things you might want to look for in a meter. Most DMMs have a capability to measure temperature with a k-type thermocouple. Some plug directly into the meter, while others plug into accessory heads that attach to the meter;

either option is good. K-type thermocouples are a good way to go. Some meters are capable of measuring capacitance. Most capacitors in HVACR are between 1μF and 600μF, so a meter capable of measuring these is helpful. Many meters do not have a capacitance range this wide so get as close as you can. The alternative to having a DMM to measure capacitance is having a standalone capacitor tester. As far as voltage, current, and resistance scales are concerned, try to select a meter that has a wide scales on all three. Voltage should be mV through about 600 VAC, current should range between a few μA to 100 A with a clamp-on ammeter accessory, and resistance should be a few ohms to 5 MΩ.

These meter specifications are only suggestions and only you can decide how your meter will be used and which features you want. Unit 3 in your textbook has many good explanations on meter settings and specifications related to safety and use.

CONTENTS

Unit 1 What You Need to Know to Understand Electricity

Unit Summary

This unit was an introductory section that introduces some basic terms used when working with HVACR equipment. It is essential to understand these basic terms in order to communicate with your supervisor, the manufacturer's tech support people, and customers. These terms will be discussed and applied in more detail in the following units. Some of these terms are easily misunderstood or confusing. Using the wrong term can change the outcome when communicating with a supervisor or other technician. Yes, there are many new terms to learn and understand. Refer back to these terms or the terms in the glossary at the end of the book to drill down to the fundamental terms needed to become a good technician.

Key Terms (Definitions can be found in the Glossary in your text.)

Alternating current (AC)	Electron
Ammeter	Hertz (Hz)
Ampere	Insulator
Circuit board	Load
Combination circuit	Ohmmeter
Complete circuit	Open circuit
Conductor	Parallel circuit
Control voltage	Resistance
Current flow	Series circuit
Cycle	Short circuit
Direct current (DC)	Volt
Disconnect	Volt meter
Electrical legend	Voltage
Electrical symbols	Watt
Electricity	Wiring diagram

Unit 2 Ohm's Law and Circuit Operation

Unit Summary

The purpose of learning Ohm's law and the watts formula is to help you understand the relationship of volts, amps, resistance, and power. Understanding this relationship will help you figure out problems and generate solutions when troubleshooting. When technicians first learn these formulas, they simply think of them as formulas that they need to learn to pass a course. These formulas, however, are the first step in figuring things out.

When you are solving problems, think of the relationships between volts, amps, resistance, and watts. Initially, you may not see the relationships. Keep trying to do so, however, because once you do, the information will become automatic and useful. You will eventually use these formulas to comprehend and decipher problems without thinking directly about the formulas themselves. Problem solving is not learned overnight, but flourishes with practice and an understanding of systems and their component parts.

Key Terms (Definitions can be found in the Glossary in your text.)

Ampere

Capacitive load

Capacitive reactance

Carbon or wire wound resistors

Combination circuits

Complete circuit

Crankcase heaters

Electric heat strips

Equal resistance method

High pressure switches

Impedance formula

Incandescent light bulbs

Inductive load

Low pressure switches

Ohm's law

Parallel circuits

Power formula

Product over sum method

Resistance

Reciprocal method

Resistive loads

Safety devices

Series circuits

Thermostats

Volt

Voltage drops

Watt

Watts formula

LAB 2.1 Application of Ohm's Law in Basic Circuits

LABORATORY OBJECTIVE

You will learn how to calculate volts, amperage, and resistance using Ohm's law.

ELECTRICITY FOR HVACR, 1e TEXT REFERENCE

Unit 2: Ohm's Law and Circuit Operation

REQUIRED MATERIALS PROVIDED BY THE STUDENT

- Calculator

- 6-in-1 screwdriver, wire cutters and wire strippers

REQUIRED MATERIALS PROVIDED BY THE SCHOOL

- 12 x 24 inch Plywood board

- Cleat receptacle to receive an incandescent light bulb

- Power cord to plug into 120 V receptacle

- Wood screws

- Wire and electrical connection components

- One 60 W and one 100 W light bulbs

SAFETY REQUIREMENTS

In this lab you will be working with 120 V so it is important that you follow all lockout/tagout procedures to prevent electrical shock and DO NOT make electrical connection with power being supplied to circuit. Do not plug in the cord without your instructor's approval. Remember to follow safe hand tool use practices and make all electrical connections neat and orderly to prevent wires from accidentally touching.

INTRODUCTION

You will calculate wattage using Ohm's law, then wire a cleat receptacle to a cord that will be plugged into a 120 V receptacle. Make sure your instructor checks your work. It is important to understand how voltage, amperage, and resistance influence each other so that you can properly size wire and breakers for a system. So we open this lab assignment with the following exercise:

Calculate the items missing in the problems below using Ohm's law and Figure 2-1-1.

Voltage: $E = I \times R$

Resistance: $R = E / I$

Current: $I = E / R$

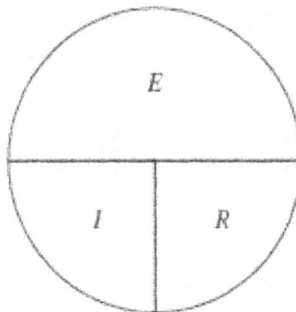

Figure 2-1-1. Use this Ohm's law pie chart to assist in using the Ohm's law formula.

1. 120 volts   6 ohms resistance   _____ amperage

2. 240 volts   6 ohms resistance   _____ amperage

3. _____ volts   8 ohms resistance   26   amperage

4. _____ volts   40 ohms resistance   6   amperage

5. 480 volts   _____ ohms resistance   26   amperage

6. 208 volts   36 ohms resistance   _____ amperage

7. 24 volts   _____ ohms resistance   3   amperage

8. _____ volts   32 ohms resistance   20   amperage

PROCEDURE

Step 1. Begin with the 60-W bulb. With your instructor's help, check the resistance of the filament inside the light bulb. Place one electrical meter probe on the silver-threaded bottom of light bulb and the other electrical meter probe on the button at the bottom of light bulb; set the meter to $R \times 1$ and take the resistance reading. What is the resistance of the 60-W bulb? Record your answer here:

Step 2. Mount the light cleat receptacle on the plywood board. With your instructor's help, wire the hot wire to the brass screw and the neutral wire to the silver screw, as shown in Figure 2-1-2. Good wiring practice is to wrap the wire around the screw in the direction that you tighten the screw.

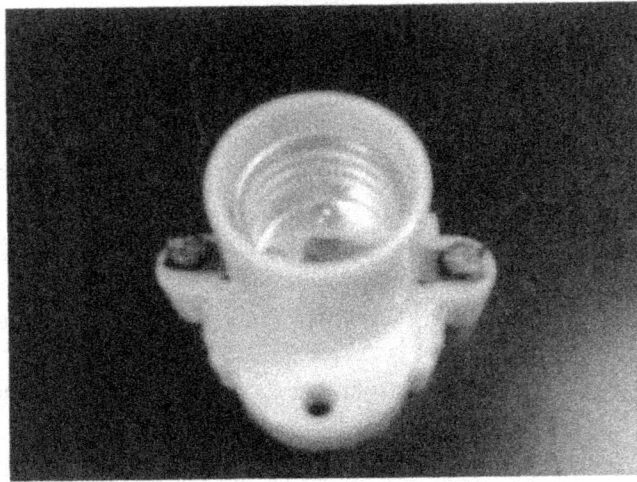

Figure 2-1-2. This shows a cleat receptacle.

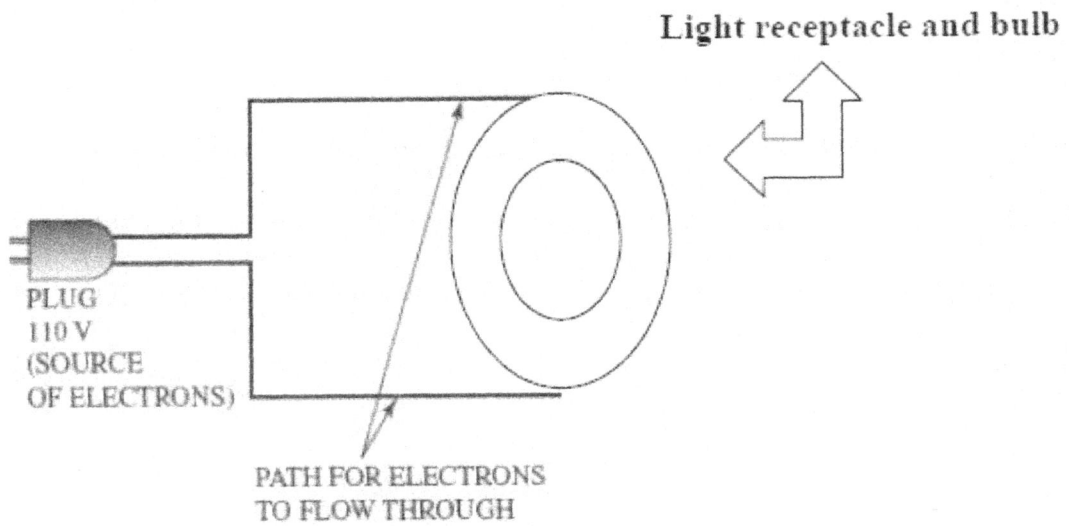

Light receptacle and bulb

PLUG
110 V
(SOURCE
OF ELECTRONS)

PATH FOR ELECTRONS
TO FLOW THROUGH

Figure 2-1-3. The light and cord assembly should be wired like this.

Step 3. Plug in the light cord with the 60-W bulb in the cleat receptacle. Did the light come on?

Step 4. It is necessary to check and record the actual voltage you have available. Voltage can vary depending on where you are. Using a digital amp meter, check and record the actual amperage of circuit here:

Calculate and record the actual wattage of the 60-W bulb here:

Step 5. Unplug the cord and remove the 60-W bulb. Repeat Step 1 and record the resistance of the 100-W bulb here:

Step 6. Plug in the light cord with the 100-W bulb in the cleat receptacle. Did the light come on?

Step 7. Repeat Step 4. Record the actual amperage of the circuit here:

Calculate and record the actual wattage of the 100-W bulb here:

QUESTIONS

1. Use the formula $W = E / I$ and the watts formula pie chart in Figure 2-1-4 to determine if the bulbs are the correct wattage. Did the actual wattage match the wattage of the bulbs tested?

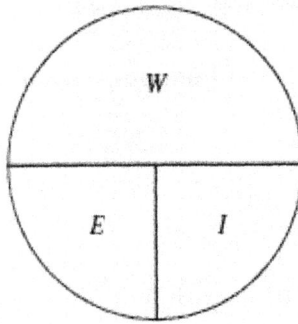

Figure 2-1-4. Use this watts formula pie chart to help determine if the bulbs are the correct wattage.

2. Which circuit had the highest amperage? Why?

LAB 2.2 Power Calculation and Ohm's Law Applied to Circuits

LABORATORY OBJECTIVE

Understand and apply series circuit laws and parallel circuit laws to an electrical circuit and to resistors.

ELECTRICITY FOR HVACR, 1e TEXT REFERENCE

Unit 2: Ohm's Law and Circuit Operation

REQUIRED MATERIALS PROVIDED BY THE STUDENT

- 6-in-1 screwdriver, wire cutter and wire strippers

- Calculator

REQUIRED MATERIALS PROVIDED BY THE SCHOOL

- 12 x 24 inch Plywood board

- Cleat receptacle to receive an incandescent light bulb

- Power cord to plug into 120 V receptacle

- Wood screws

- Wire and electrical connection components

- Two 60 W and two 100 W light bulbs

- Three resisters of known value

- Electric meter

SAFETY REQUIREMENTS

In this lab you will be working with 120 V so it is important that you follow all lockout/tagout procedures to prevent electrical shock and DO NOT make electrical connection with power being supplied to circuit. Do not plug in cord without instructor approval. Remember to follow all safe hand-tool use practices. Make all electrical connections neat and orderly to prevent wires from accidentally touching.

INTRODUCTION

You will wire two cleat receptacles in series and answer questions pertaining to the wire project. The purpose of this Lab is to help you understand the difference between series circuits and parallel circuits and how voltage, current, and resistance influence each other differently in series and parallel circuits. You will wire two cleat receptacles in series and two cleat receptacles in parallel and answer questions pertaining to the project. Lastly, you will use resistors to calculate total resistance for series and parallel circuits.

Components of circuits can be wired in two different ways, in series or in parallel with other devices. Loads are almost always wired in parallel with other loads and switches are always wired in series with the loads that they control. Most circuits in HVACR are a combination of these two wiring schemes; loads are in parallel with other loads and the switches that control these loads will be wired in series with them, thus the term combination circuits.

For the purposes of demonstration the cleats will be first wired in series, although this is unusual among loads in HVACR systems and it is important to note that this is not the normal way of wiring loads. Series circuits follow Ohm's law according to the following formulas, where E is voltage, I is current, and R is resistance:

$$E_t = E_1 + E_2 + E_3 + E_4 \ldots$$

$$I_t = I_1 = I_2 = I_3 = I_4 \ldots$$

$$R_t = R_1 + R_2 + R_3 + R_4 \ldots$$

As mentioned, most loads are wired in parallel with other loads. They are normally wired this way so that the fully applied voltage is available to each load. HVACR components, especially motors, do not deal well with less than their rated voltage. Loads wired in parallel follow Ohm's law according to the following formulas:

$$E_t = E_1 = E_2 = E_3 = E_4 \ldots$$

$$I_t = I_1 + I_2 + I_3 + I_4 \ldots$$

$$\frac{1}{R_t} = \frac{1}{R_1} + \frac{1}{R_2} + \frac{1}{R_3} + \frac{1}{R_4} \ldots$$

PROCEDURE PART I: Series Circuit

Step 1. Mount two cleat receptacles, like the one shown in Figure 2-2-1, on plywood.

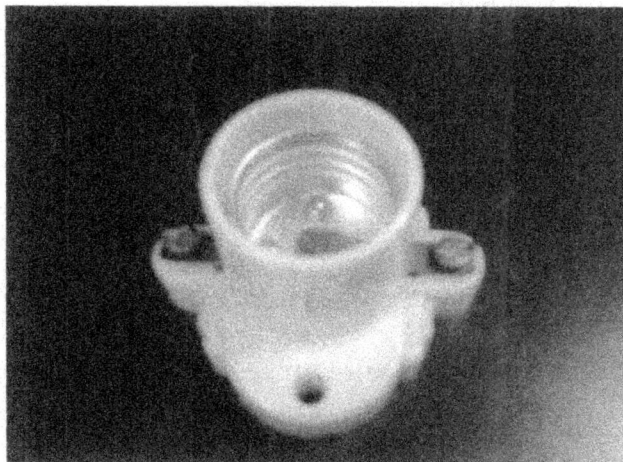

Figure 2-2-1. Cleat receptacle

Wire the cleats in series like the diagram shown in Figure 2-2-2. Have your instructor check your wiring.

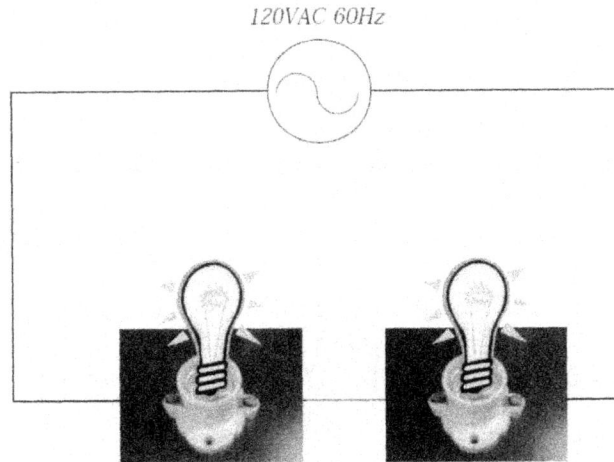

120VAC 60Hz

Figure 2-2-2. This diagram shows the cleats wired in series.

Step 2. Install two 60 W bulbs in cleats and plug into 120 VAC.

Do both lights burn with equal brightness? _____

Check and record the amperage before the first bulb, between the bulbs, and after the second bulb.

Amperage before the first bulb _____

Amperage between the bulbs _____

Amperage after the second bulb _____

Is the amperage the same in all places checked? _____

Step 3. Unscrew one bulb. What happens to the other light? _____

Replace one bulb with a 100 W bulb and plug in. Which bulb burns the brightest?

_____

Check and record the amperage before the first bulb, between the bulbs, and after the second bulb.

Amperage before the first bulb _____

Amperage between the bulbs _____

Amperage after the second bulb _____

Is the amperage the same in all places checked? _____

Unscrew one bulb. What happens to the other? _____

PROCEDURE PART II: Parallel Circuits

Step 4. Wire the cleat receptacles in parallel like the diagram in Figure 2-2-3. Have your instructor check your wiring.

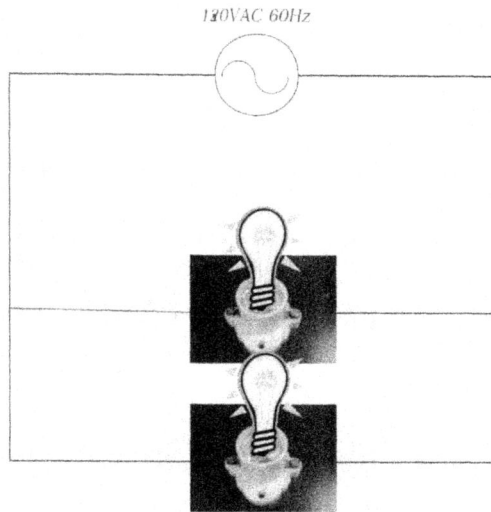

Figure 2-2-3. This diagram shows the cleats wired in parallel.

Step 5. Install two 100 W bulbs into the cleats and plug into the 120 V receptacle.

Do both lights burn with equal brightness? _____

Check and record the amperage before the first bulb and at each bulb.

Amperage before the first bulb _____

Amperage between bulbs _____

Total amperage _____

Step 6. Unscrew one bulb, it does not matter which one.

What happens to the other light? _____

Explain why.

Step 7. Replace one bulb with a 60 W bulb (it does not matter which one) and plug in.

Is one light brighter than the other? _____

Check and record the amperage before the first bulb and between the bulbs.

Amperage of the 100 W bulb _____

Amperage of the 60 W bulb _____

Total amperage _____

PROCEDURE PART III: Resistors in Series and in Parallel

Acquire three resistors (of known value), like the ones shown in Figure 2-2-4. You will connect them first in series, then in parallel. DO NOT ENERGIZE the resistors. This is to be done without power applied.

Step 8. Measure the resistance of each resistor by setting the ohmmeter to R x 1, place one meter lead on one end and the other meter lead on the other end. Be careful not to touch the meter leads with fingers as that will check the resistance of your body.

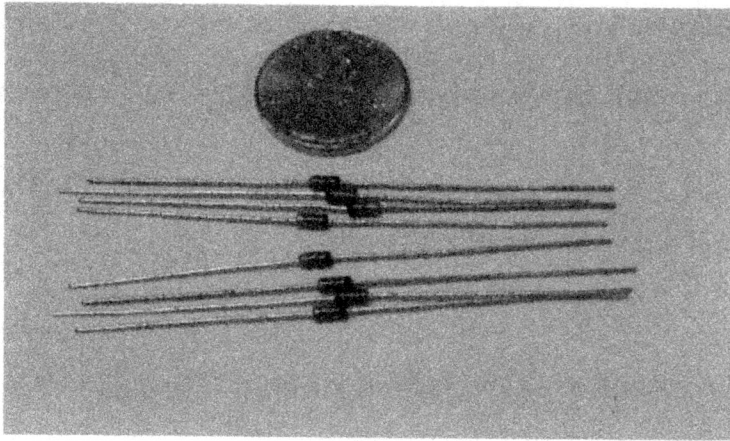

Figure 2-2-4. This is an example of resistors. The resistors your instructor assigned to you may be larger and have different colored bands.

R_1 _____ R_2 _____ R_3 _____

Calculate R_t (resistance total) from the resistors above.

R_t = _____

Step 9. Now twist them together end to end so that all are in series. Measure the actual

resistance using the ohmmeter.

Actual resistance _____

Measure the resistance of each resistor by setting ohmmeter to $R \times 1$, place one meter lead on

one end and the other meter lead on the other end. Be careful not to touch the meter leads with

fingers as that will check the resistance of your body.

R_1 _____ R_2 _____ R_3 _____

Calculate R_t (resistance total) from the resistors above.

R_t = _____

Step 10. Twist all three resistors together at one end, then twist all three resistors together at the other end so all are in parallel, each having its own path. Measure the actual resistance with ohmmeter.

Check and record the actual resistance of the circuit _____

QUESTIONS

1. Complete the Ohm's law calculations below for a parallel circuit (refer to the discussion in the introduction to this lab). Use the following information: incoming voltage is 115 VAC; R_1 resistor is 15 ohms; R_2 resistor is 5 ohms; and R_3 resistor is 12 ohms.

E_t 115 E_1 _____ E_2 _____ E_3 _____

I_t _____ I_1 _____ I_2 _____ I_3 _____

Rt _____ R_1 15 R_2 5 R_3 12

2. Complete the Ohm's law calculations below for a parallel circuit (refer to the discussion in the introduction to this lab). Use the following information: incoming voltage is 240 VAC; R_1 resistor is 25 ohms; R_2 resistor is 14 ohms; and R_3 resistor is 18 ohms.

E_t 240 E_1 _____ E_2 _____ E_3 _____

I_t _____ I_1 _____ I_2 _____ I_3 _____

Rt _____ R_1 25 R_2 14 R_3 18

3. The electrical loads of an HVACR unit are wired in _____.

    a. Series   b. Parallel

4. Switches in a HVACR unit are wired in _____.

    a. Series   b. Parallel

5. If a switch in series with an electrical load opens, what will happen to the load?

    a. Turn off   b. Keep running

6. The sum of all voltage drops for a series circuit should total the incoming voltage.

    a. True   b. False

7. The voltage to a load in a parallel circuit is deducted from the total voltage.

    a. True   b. False

Unit 3 Safe Use of Electrical Instruments

Unit Summary

This unit discussed the features and benefits of digital multimeters and clamp-on ammeters. It is important to select electrical instruments based on their safety, quality, and the features it has that you require in order to do a good job. Price is usually the first consideration that influences a technician's decision to purchase a meter—but price should be the last consideration. Consider what you pay for a quality instrument on an annual basis and you will notice that the yearly cost is not that much. A good quality meter will last more than 10 years. Divide the cost of the meter by 10 to determine the annual cost of having a safe and accurate instrument to use. For example, if you purchase a DMM with a life expectancy of 10 years for $400, that would equal an investment of $40 per year or about $3 a month. Most technicians would be willing to rent a quality meter for $3 a month. *Remember:* These are your professional tools and you need them to do a good job.

Key Terms (Definitions can be found in the Glossary in your text.)

Analog multimeter

Arc flash

Auto feature

Clamp-on ammeter

Digital multimeter

Ghost voltage

Lockout/tagout

Milliammeter

National Fire Protection Association

Non-contact voltage detectors

Ohmmeter

Personal protection equipment

Range

Resolution

Root means square

Symbols

Temperature tester

Voltmeter

LAB 3.1 Basic Multimeter and VOM Resistance Measurements

LABORATORY OBJECTIVE

After completion of this lab activity you will be able to take resistance measurements of any type of electrical load competently and accurately. You will also be able to use electrical notation to describe resistance and voltage and current values more concisely.

ELECTRICITY FOR HVACR, 1e TEXT REFERENCE

Unit 3: Safe Use of Electrical Instruments

Unit 18: Resistors

REQUIRED MATERIALS PROVIDED BY THE STUDENT

- DMM or VOM suitable for HVACR field work

  (Note: Unit 3 of the text describes how to select an appropriate meter)

REQUIRED MATERIALS PROVIDED BY THE SCHOOL

- A supply of six numbered resistors with various resistances

- A supply of six numbered electrical loads

SAFETY REQUIREMENTS

Since we are not dealing with any "live" electricity for this laboratory activity there is no risk of electric shock. Some safety precautions do apply, though. As with any laboratory activity, safety glasses should be worn at all times. Before attempting to use any DMM or VOM make sure you read the instructions and know the limits of your meter.

INTRODUCTION

For this laboratory exercise we will be examining several components typically found in electrical circuits and taking resistance readings, which are measured in ohms. There are two general rules to follow before making any resistance measurement: (1) always de-energize the device being tested first as you should never have power applied to any load when taking a resistance measurement; and (2) you should isolate the device you are testing from the rest of the circuit so you do not measure the circuit connected to the device.

When troubleshooting a circuit, it's important to know what you need to measure and what reading to expect. In addition, before taking the measurement you should know how to set up the meter range, scale, and leads for taking that specific measurement.

There are two general categories of meters used in the field: manual versus auto-ranging meters and digital versus analog meters. Auto-ranging meters select the most appropriate scales automatically under most circumstances, but not always. With manual ranging meters the user has to select the appropriate scale, which can be an advantage. For the purposes of this lab it would be most beneficial to the student to turn off the auto-range feature on your meter. Digital multi-meters are the tool of choice for most HVACR technicians, although analog meters can also be used. The differences in meters can be understood completely by reading Unit 3 of your textbook.

Electrical notation is important for this lab and those that follow. Notation is a way of shortening what would otherwise be a very long number. Table 3-1-1 shows the rules when determining what a notation symbol means when it shows up on a meter reading.

Symbol	Multiplier	Example
μ	Means: $^1/_{1,000,000}$ (Shift decimal to the left 6 times or divide by 1,000,000)	15.0μA = 0.000015A
m	Means: $^1/_{1,000}$ (Shift decimal to the left 3 times or divide by 1,000)	3.2mA = 0.0032A
k	Means: 1,000 (Shift decimal to the right 3 times or multiply by 1,000)	1.85kΩ = 1,850Ω
M	Means: 1,000,000 (Shift decimal to the right 6 times or multiply by 1,000,000)	34.5MΩ = 34,500,000Ω

Table 3-1-1. Shows common notation symbols used in HVACR electrical measurements and how they can be converted.

Once you have some resistance reading (other than infinity) at the highest meter scale, it's important to know if your meter will give you a good reading at the next lower scale. If you select a voltage or current scale lower than the value you are trying to measure you can damage the meter. However, you should get to the lowest scale your meter will allow to get the most accurate reading. Figure 3-1-1 shows a VOM set for a 40kμ scale and is indicating 410Ω. Figure

3-1-2 is showing the same measurement except the scale is set at 4kΩ, and the meter is indicating 420Ω, which is a more accurate reading than the first measurement. Figure 3-1-3 shows the meter trying to read the resistor in the 400Ω range but since the resistor is higher than 400Ω it shows infinity; the scale was turned down too far.

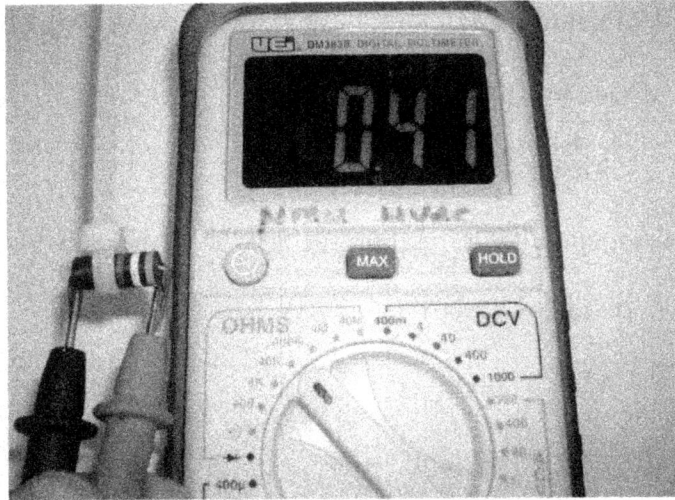

Figure 3-1-1. This is a VOM indicating a reading of 410 ohms at the 40kΩ scale.

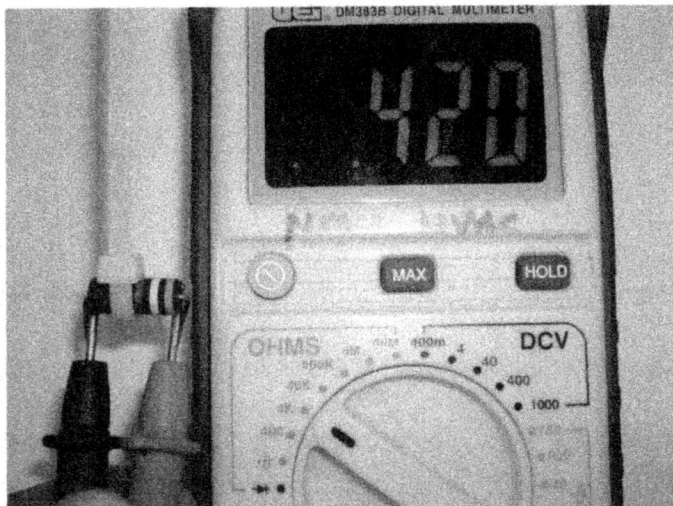

Figure 3-1-2. This is a VOM indicating a reading of 420 ohms at the 4kΩ scale. This is a more accurate reading than the first measurement. Can we go down to the 400Ω scale and get a good reading?

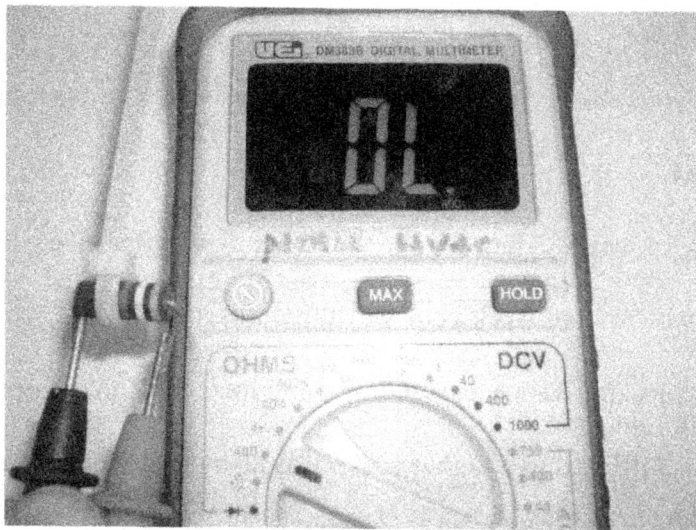

Figure 3-1-3. This time we shifted the scale to 400Ω but the meter shows infinity. This is because the resistor has a higher value than the scale the meter is set for.

Your meter will always indicate one of three readings:

1. An infinite amount of resistance, which is usually indicated as "OL" or sometimes "1." Be careful you do not interpret the "1." as 1 ohm. To find out how your meter indicates infinity, set your meter to any resistance range and hold the two probes apart from one another, the meter will be showing its version of infinity.

2. Zero resistance is usually indicated as "0.00" or a very small number that is close to 0. You would find little to no resistance across the contacts of a switch if it was closed or "on."

3. Measurable resistance anywhere between infinity and 0 can be observed when measuring the resistance of any load such as a motor or coil.

The resistances of different load types vary greatly. Some loads, like a compressor motor windings, may measure only a few ohms, while other loads, like some relay coils, can measure in thousands of ohms.

PROCEDURE

Step 1. Using the resistors received from your instructor, measure the resistance of each one.

Helpful Resistance Measuring Hints:

- Generally for measuring an unknown resistance value you want to set your meter for the highest range the meter is capable of and work your way down the scales.

- If the meter indicates infinity (usually shown as "OL" or sometimes "1.") then you may have the scale set too low; try shifting it to a higher scale. It is also possible that the meter is not capable of reading a resistance as high as the one you are trying to measure, or that the device you're trying to measure is open.

- If the reading on your meter bounces back and forth and does not settle on a number after a few seconds, your test probe tips are probably not making good contact with the device.

- You should only touch one (or no) test probe tip with your fingers when taking a measurement. If you are reading a high resistance device and touch both probes with your skin, the meter could actually be showing the resistance of your body. (It depends on how much iron was in your cereal this morning!)

Record the measured value of the resistors in the table below:

Resistor Number	Measured Value (Ω)
1	
2	
3	
4	
5	
6	

Step 2. Using Figure 18-4 found in your textbook, find the value of each of the six resistors using the color code. Compare values you measured from those that you calculated using the chart. If they are not the close to the same, re-check the resistor using a meter to determine if you made an error.

Record the calculated value of the resistors in the table below:

Resistor Number	Calculated Value (Ω)
1	
2	
3	
4	
5	
6	

Step 3. Using the six numbered electrical loads provided, measure the resistance of each.

Record the measured value of the loads in the table below:

Load Number	Measured Value (Ω)
1	
2	
3	
4	
5	
6	

QUESTIONS

1. Which of the three meter readings should be found when measuring the resistance of any electrical load: open (infinity), closed (0), or some measurable resistance between infinity and 0?

2. What are the three electrical notation symbols commonly used in HVACR work? Why are these symbols useful?

3. What are the two ways a DMM or VOM can show an infinite amount of resistance?

4. You are measuring the resistance of a load and the meter is in the "20MΩ" scale. The display reads "0.33." How many ohms is the load?

5. You are measuring the resistance of a load and the meter is in the "2kΩ" scale. The display reads "1.56." How many ohms is the load?

6. You are measuring the resistance of another load and the meter is in the "20MΩ" scale. The display reads "0.03." Can you switch the meter down a scale if the next lower option is "20kΩ"? What if the next lower meter scale is "40kΩ"?

LAB 3.2 Taking Voltage and Current Measurements

LABORATORY OBJECTIVE

After completion of this lab activity you will be able to take voltage and current measurements of any type of electrical load safely and accurately. You should also be able to use electrical notation to describe voltage and current values more concisely.

ELECTRICITY FOR HVACR, 1e TEXT REFERENCE

Unit 3: Safe Use of Electrical Instruments

REQUIRED MATERIALS PROVIDED BY THE STUDENT

* DMM or VOM suitable for HVACR field work (Note: Unit 3 of the text describes how to select an appropriate meter)

REQUIRED MATERIALS PROVIDED BY THE SCHOOL

* A basic wiring trainer

SAFETY REQUIREMENTS

Voltage and current measurements must be taken with power applied to a circuit, so we will be working with "live" circuits and components. Do not plug in the circuit until you have received the approval of your instructor. Remember to follow safe hand-tool use practices and make all electrical connections neat and orderly to prevent wires from accidentally touching. Lockout/tagout procedures should also be followed. When possible you should turn off the power supply before connecting the meter to take a reading, then you can turn the power back on. You should know how to quickly turn power off to the circuit in the event of a problem, which is usually by pulling out a plug or turning off a main switch. You should review Unit 3 in your textbook before attempting this lab assignment to make sure you know how to operate your meter when taking voltage and current measurements.

INTRODUCTION

Voltage

For this lab we will be measuring alternating current power supplies (AC), which will be abbreviated VAC, ACV, or V~ on your VOM. Voltage measurements will usually be nominal values such as 24 VAC, 120 VAC, 208 VAC, and 230 VAC. There will be some variations in your measurements. For example, a power source might provide a nominal 120 VAC, but your actual measurement may vary slightly, perhaps 116 VAC or 123 VAC. Normally, voltage should be within +/- 10% of the nominal voltage to be acceptable.

When taking measurements, you must remember DMM's and VOM's will indicate a <u>difference</u> in voltage. If you measure from a point on a hot conductor to another point on the same hot conductor your reading will be 0 V. This will be important to remember when we use voltage to test for open switches or other breaks in a circuit.

Current

Current, which is essentially electrical volume, will also be measured on alternating power supplies for this lab. Current is measured in amps (or amperes). Abbreviations on your VOM are usually ACA or A~. The amount of current is dependent on the voltage of the power source and the type and size of load that is being powered. All loads will draw some amount of current. As it is with voltage, you should always have an idea of what current to expect before taking a reading.

There are two ways meters can be used to measure current. The first method is known as the "meter in series" method. Imagine the meter like a wire: current that flows through the load has to flow through the meter as well. It is important to not exceed the current rating of your DMM, you blow in internal meter fuse if you do. Most meters have a limit of 10 or 20 amps when measuring current this way. Make sure you do not short out the meter when attaching your probes. Do not attempt to bypass a load with the DMM set for current. The meter is essentially a wire; going around any load with a wire is a short circuit.

Measuring current with a clamp-on meter is the other method. DMM's that are set up for this either have a clamp at the end of the meter or one can be attached with wires. The clamp-on method is usually the method of choice in HVACR work since it is fairly accurate and you can take much higher readings than by using the meter in series method. Normally, a DMM or VOM will be designed to measure current in one of these ways and cannot measure both ways. See Figure 3-2-1 for observing current measurements with both types of meters.

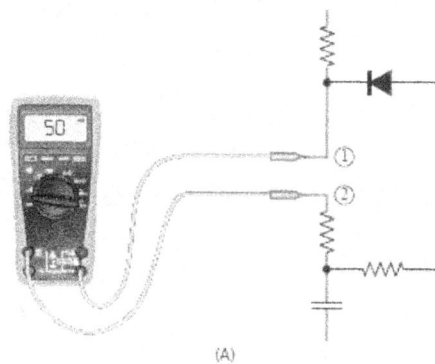

(A)

Figure 3-2-1. This shows the clamp on method of current measurement (left) and meter in series method (right).

PROCEDURE

Step 1. With the circuit wiring trainer provided, wire the circuit found in Figure 3-2-2. Check with your instructor after completing it.

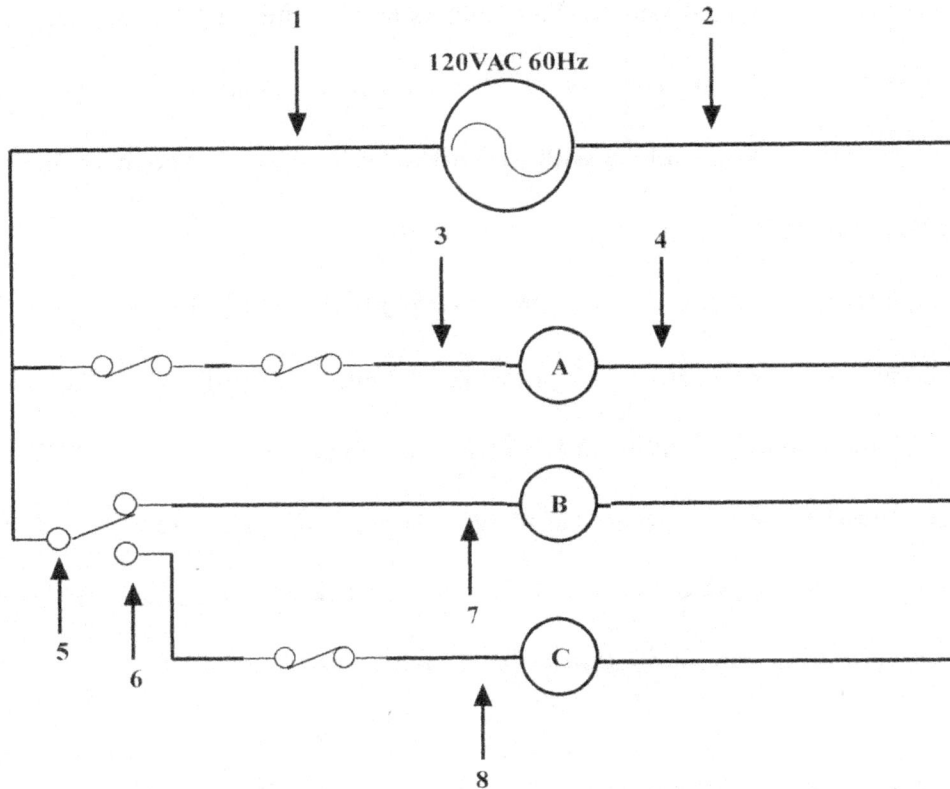

Figure 3-2-2. Use this diagram to complete your wiring task for step 1.

Step 2. You should have some expectation of what voltage will be before a measurement is taken. Approximately what voltage would you expect between points 1 and 2 on the diagram? Record this value below.

_____VAC

Step 3. With the power on, measure the voltage of the power source using at points 1 and 2 on the diagram and record below.

_____VAC

Step 4. What voltage would you expect to measure between points 3 and 4 on the diagram? Record this value below.

_____VAC

Step 5. Measure the voltage across load A (while it is energized) at points 3 and 4 on the diagram and record below.

_____VAC

Step 6. If the single pole double throw (SPDT) switch in the diagram is in the top position (as shown), what would you get for a voltage reading from points 5 and 6? Record your estimate.

_____VAC

Step 7. Measure the voltage between points 5 and 6 on the diagram with your meter and record it below.

_____VAC

Step 8. While load A and either B _or_ C is on, measure the current at points 3, 7, and 8 on the diagram and record the three values below.

Point 3_____ AAC

Point 7 _____ AAC

Point 8_____ AAC

Step 9. With the switches in the same position as they were for step 5, what current would you expect if you measured current at point 1 or 2 on the diagram? Record this value below.

_____ AAC

Step 10. Measure the current at points 1 and 2 from the diagram and see if it matches your prediction.

_____AAC

QUESTIONS

1. In industry lingo it is said an electrical load will _____ amps or current.

2. In Step 3 you measured voltage between points 1 and 2 on the diagram. Would you get the same reading or a different one if you measured between points 1 and 4?

3. What do you think it would mean if you measured a voltage of "50 VAC" between points 1 and 2?

4. You have discussed combination circuits in previous assignments. Using what you know estimate what the total current would be from the power source if we could energize loads A, B and C all at the same time.

5. If the switches were arranged so loads A and C were energized and load B was de-energized, list the arrow numbers where you could measure the current through only load C.

6. In Step 5 you measured the voltage across load A and likely got about 120 VAC. What devices might you suspect are problems if the voltage was measured at 0 VAC but load A was supposed to be on? How would you identify which one is the problem?

7. What are the two ways of measuring current? Which one would be best for measuring current that you expect to be fairly high (above 20 amps)?

8. What would you expect to see happen to the total current if more loads were added in parallel to the others in this or in any circuit?

9. Convert 45 mA into amps.

10. The symbol "V~" on a DMM or VOM indicates what value?

11. The symbol "AAC" on a DMM or VOM indicates what value?

12. Why is it important for a technician to have an estimate of voltage or current before actually taking the reading?

Unit 4 Electrical Fasteners

Unit Summary

Knowing the names of fasteners and how they work is a valuable tool for all HVACR professionals. Technicians should know as much as they can about everything so that they will be a valuable asset to a company and to themselves. The HVACR professional tends to be a jack of all trades and a master of HVACR. The true professional must know air conditioning, heating, and refrigeration. He or she must also know electrical work, plumbing, and some carpentry. Our career calls for a well-rounded, skilled person, or at least for a person who is willing to learn skills outside of the narrow HVACR skill set.

Knowing about the various types of fasteners will help you complete jobs in a quick and efficient manner. For example, quick connectors speed up the processes of checking and changing out components. Knowing what fasteners are available, how to use them, and calling them by their correct name makes the technician a professional. You will find that there are many different terms for the fasteners we discussed in this unit. When discussing fasteners it is important for you to be on the same page as the person with whom you are communicating.

Key Terms (Definitions can be found in the Glossary in your text.)

Bolt	Gauge number	Quick connectors	Sheet metal screw
Butt connector	Hex head	Right-hand thread	Sheetrock screw
Carriage bolts	Hex screwdriver	Ring connector	Underwriter's Listing
Crimping tool	Insulation stripping tool	Round-tipped screw	Wire cutter
CSA Standard	J-bolts	SAE J429	Wire gauge
Double male, single female connector	Machine screws	Screw	Wire nuts
	Magnetic chuck	Self-drilling	Wood screws
Electrical tape	Male slip-on connector	Self-starting screw	
Fastener	Nails	Self-tapping screw	
Female slip-on connector	Nut drivers	Self-threading screw	
Connector	Nylon strap	Set screws	
Flag connector	Nylon ties		
Fork connector	Penny		

LAB 4.1 Electrical Connections

LABORATORY OBJECTIVE

After completing this lab you should understand the importance of making proper electrical connections to ensure safety and equipment longevity. You should also know how to select the correct type of connection for all applications and assemble them properly.

ELECTRICITY FOR HVACR, 1e TEXT REFERENCE

Unit 4: Electrical Fasteners

REQUIRED MATERIALS PROVIDED BY THE STUDENT

- 6-in-1 screwdriver (or set of screwdrivers)

- Wire strippers/crimpers/cutters

REQUIRED MATERIALS PROVIDED BY THE SCHOOL

- A selection of wires in various gauges (THHN/THWN, and SJ cord)

- Male plug and Female cord receptacle

- Wire nuts, quick connect (female spade) terminals, ring connectors, lug/pressure plate connectors, and butt connectors adequate for wire gauge sizes used

SAFETY REQUIREMENTS

Since we are not dealing with any "live" electricity for this laboratory activity there is no risk of electric shock. Some safety precautions do apply though. As with any laboratory activity safety glasses should be worn at all times. Gloves would be a good idea as well since working with copper wires can be hard on your hands.

INTRODUCTION

There are many types of electrical connections used in HVACR and it is beneficial for you to know the names of each and how they are used. The importance of good electrical connections cannot be overstated. Poor connections can cause several problems. The first one is the risk of overheating connectors or wires and starting a fire. Any connection that is not made properly can have some resistance built in. If the connection is made so that wires intermittently touch inside the fastener then sparking can occur inside the connector, which can cause fires as well as damage to equipment caused by this rapid arcing. Poor connections containing a resistance can also cause equipment problems because resistance means that there will be a voltage drop. Some types of equipment are very sensitive to low voltage and components can be damaged, especially systems that contain electronic devices with printed circuit boards. In all cases, good contact between conductors and fasteners is key.

PROCEDURE

Helpful Connection Making Hints

- Always use a crimping tool to crimp quick connect (spade) terminals, ring terminals, or butt connectors to a stripped wire, never use pliers or channel locks (see Figure 4-1-1).

- When making a wire nut connection, don't pre-twist the wires; let the turning wire nut twist them together (see Figure 4-1-2).

- A completed connection should NEVER show the copper or aluminum conductor protruding from the bottom of the connector (see Figure 4-1-3).

- For any type of connection, the wires should not pull out of the connector without a substantial amount of force. If they easily pull apart the connection was not strong enough.

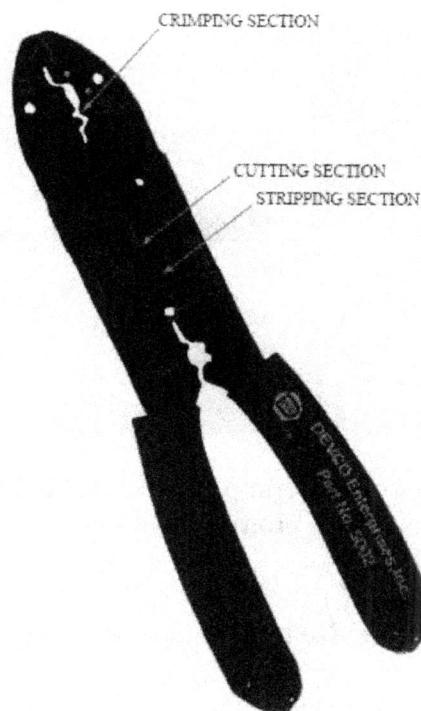

CRIMPING SECTION

CUTTING SECTION
STRIPPING SECTION

Figure 4-1-1. This is a wire cutting, stripping, and crimping tool that should be used when making crimped connections.

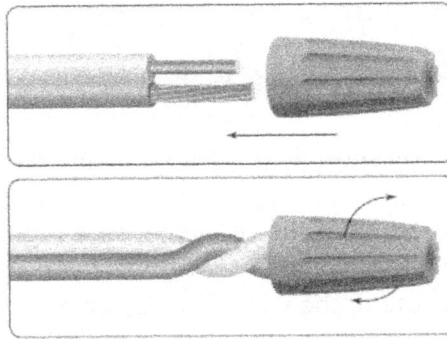

Figure 4-1-2. This shows the correct way to make a wire nut connection, let the nut do the twisting. Wire should be stripped $^7/_{16}$" for 16 AWG wires and $^3/_8$" for all other sizes.

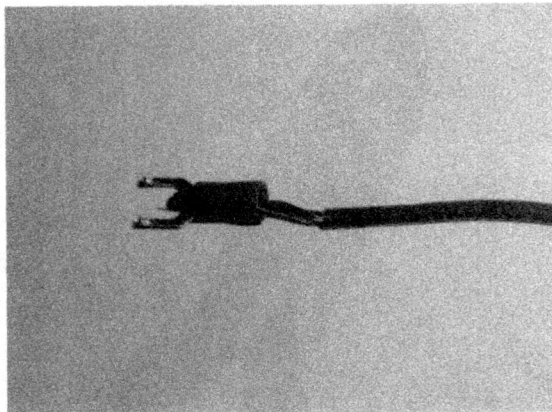

Figure 4-1-3. This shows what a crimp connector should NOT look like when finished. Notice how the copper conductor protrudes out the bottom of this forked connector.

Step 1. Use a wire nut of the correct size to fasten together two wires of the same gauge. (See Tables 4-1 and 4-2 in your textbook to determine the correct size wire nut for your application.) Show it to your instructor for approval.

Step 2. Use a wire nut of the correct size to fasten together two wires of DIFFERENT gauge sizes. Show it to your instructor for approval.

Step 3. Using a stripped piece of wire and the correct size of quick connect terminal, crimp it on to the wire using the correct tool. Attach this to a female spade connection provided by your instructor. Show it to your instructor for approval.

Step 4. Take two stripped pieces of the same gauge size wire and use a butt connector to attach them together with the correct tool. Show it to your instructor for approval.

Step 5. Using another stripped wire, attach a forked terminal connector to it and attach it to a terminal strip provided by your instructor. Show it to your instructor for approval.

Step 6. Using another piece of stripped wire, attach the correct size ring terminal to it. Fasten this to a ring terminal strip or device provided by your instructor. Show it to your instructor for approval.

Step 7. Using another piece of stripped wire, attach it to a lug type connection provided by your instructor. Show it to your instructor for approval.

Step 8. Using the supplied 115V male plug and female cord end receptacle and a short piece of type SJ cord, attach both plugs to the SJ cord properly. Be careful not to nick the insulation inside of the individual wires when removing the outer jacket insulation. Make sure to follow the correct polarity of the plug (gold terminal = hot, silver terminal = neutral, green terminal = ground). Show it to your instructor for approval.

QUESTIONS

1. What is the difference between a forked connector and a ring connector? What do you think are the advantages of each?

2. Name a reason why you might have to install a male cord end or female receptacle in HVACR work.

3. What are the two major concerns when it comes to poor electrical connections?

4. How many 14 AWG conductors can you install in a yellow color wire nut?

5. You should only use what tool for crimp on type connections?

Unit 5 Power Distribution

Unit Summary

The information in this unit may not seem relevant to the HVACR profession, but technicians must know how power is supplied to a building. Connecting the wrong voltage source may damage the equipment.

The power supplies discussed in this unit are not the only electrical sources provided to our customers, but these are the most common types. Single-phase power is used with residential homes and apartments. Various three-phase power options are used in commercial and industrial installations.

Common voltages in residential buildings are single-phase 208 or 240 volts. Commercial structures use three-phase voltage nominally rated at 208, 240, 440, or 480 volts. The voltage supplied by the power company should be within ±10% of the rated voltage. Low- or high-voltage conditions can create equipment damage.

It is important to know the power distribution output. Understanding the supply voltage is the first step in basic troubleshooting.

Key Terms (Definitions can be found in the Glossary in your text.)

Connected load	Overhead wiring
Delta transformer	Power
Entrance panel	Power distribution
Generator	Power factor
High leg	Power generation
Kilovolt-amps	Service drop
Kilowatt	Transmission lines
Kilowatt hour	Underground wiring
Megawatt	Watt
Megawatt hour	Wye transformer
Meter loop	

Unit 6 National Electric Code®

Unit Summary

The purpose of this unit was to familiarize you with the National Electrical Code and how it impacts our profession. The NEC is a safety code that covers electrical installations and electrical components. It is important to understand that although HVACR has a code called the Uniform Mechanical Code, we must also follow all building, plumbing, electrical, and life safety codes.

This unit discussed several electrical requirements that must be considered when installing HVACR equipment. For example:

- Required disconnects (unless the breaker can is nearby and can be seen by the technician)
- Wire sizing
- Use of nameplate data to size wire and breakers
- Electrical requirements directed by the NEC.

All HVACR equipment shall have a positive means of removing or disconnecting power. The electrical disconnect is generally sized to handle 125% of the rated load amperage.

General wiring in commercial and industrial buildings shall be protected inside conduit. Open wiring, meaning wiring not inside conduit, is allowed inside residential installations. Even if not required in many residential installations, it is a good idea to have all conductors in conduit when passing through supply or return air ducts. Conductor insulation burning inside a duct can quickly distribute smoke throughout a structure.

NEC Table 310.15(B)(16) is one of the most important sections in the NEC. Table 310.15(B)(16) is used by technicians to determine the correct wire size based on conductor type (copper or aluminum) and application temperature. If a wire burns open or overheats, the wire table should be consulted. Burned or overheated wires are not always caused by undersized conductors. This condition can also be caused by loose connections or a load that is partially shorted. An asterisk (*) on wire sizes in NEC Table 310.15(B)(16) means that the maximum wire size should be reduced by 5 A. This exception affects wire sizes AWG 14, 12, and 10.

We learned that there is important information on wire insulation. The cover or jacket of the wire shows the AWG wire size, and the letters on the insulation indicate the features of the insulation coating. THHN and THWN wire insulation is commonly used when hooking up HVACR equipment. Understanding and using the NEC is extremely important for the advanced technician.

Key Terms (Definitions can be found in the Glossary in your text.)

Air handler	Branch circuit
American Wire Gauge	Code jurisdiction
Ampacity	Conductor
Article	Copper clad aluminum conductor

Disconnect

Duct heater

Fuse sizes

Grounding

Hermetic compressors

Listed

Lockout/tagout

Locked-rotor amps

National Electrical Code (NEC)®

National Fire Protection Association

NEC Article 440

NFPA 70®

Nonmetallic sheathed cable

Overcurrent protection

Rated load current

Readily accessible

Romex

Self-contained electric heating

Wire gauge

LAB 6.1 National Electric Code® Considerations

LABORATORY OBJECTIVE

You will understand how the National Electric Code (NEC) can assist you in safe electrical HVACR installations by proper wire sizing and fuse selection.

***ELECTRICITY FOR HVACR, 1e* TEXT REFERENCE**

Unit 6: National Electric Code®

REQUIRED MATERIALS PROVIDED BY THE STUDENT

- None

REQUIRED MATERIALS PROVIDED BY THE SCHOOL

- 6-in-1 screw driver

- Five HVACR units with nameplates

- Five electric motors used in the HVACR industry

- Five different wire sizes of any length

SAFETY REQUIREMENTS

Care should be taken when taking off unit covers to prevent hand injury. Always remember: if you take a cover off, put it back on. Power should be OFF to all equipment so follow all lockout/tagout procedures if units are in working order. Always follow all lab safety rules.

INTRODUCTION

This lab is designed to get you acquainted with reading the information found on HVACR system nameplates. These nameplates are full of useful information for service, installation, and troubleshooting. Two pieces of particularly useful data is the various amperage draw numbers of the unit and the individual components inside. This information, along with the allowable ampacities table of the National Electric Code (NEC), are the tools HVACR installers use to correctly size fuses or breakers and wire.

Typically, larger systems will draw greater amounts of current and will require larger gauge conductors, but it really depends on the type of system and the voltage at which it operates. Amperage is only one of the considerations of sizing a circuit breaker, fuse, or disconnect for the equipment; the voltage also plays a role. The 120 VAC power supplies that are typical of window air conditioner units and some other packaged systems require the use of a single pole circuit breaker or single fuse in the hot conductor. It is an NEC code violation to install any switch, fuse, or circuit breaker that breaks a neutral wire. The 120 VAC power sources use two conductors: one hot and one neutral. The 208 or 230 VAC power sources also have two conductors, but they are both hot. That means a two-pole breaker or two fuses (one for each hot) must be used. Figure 6-1-1 shows a single pole circuit breaker and a two-pole circuit breaker for side-by-side comparison. Three-phase systems will use three-pole circuit breakers or three fuses to protect the circuit. If it was installed properly, you can identify the voltage of any system simply by looking at the breaker or fuse protection it has.

The supply power wires sometimes must be protected from external conditions such as weather, sunlight, or damage from abrasion or movement. There are many different types of electrical conduits that are suited for every application you can think of. Popular types include watertight conduits, galvanized steel conduits (also called thinwall or EMT), and flexible aluminum clad conduits (also called MC conduits). Your conditions will dictate which, if any, wire protection is required.

The NEC is a national code that applies to all areas of the country. Local codes have to be at least as stringent, but can have additional requirements that go over and above those of the NEC. It is always best to consult with a local electrical code department to determine the exact requirements in your area. There are many other NEC considerations, but many of them will be handled by a qualified electrician who will normally install most of the power sources and conductors for a system. HVACR technicians in most jurisdictions are responsible for internal unit wiring and control wiring. Again, if you are unsure of what you can lawfully work on or install you should contact a local electrical code official. The rules we cover in this lab will give you a good start in understanding the role NEC plays in protecting people, buildings, and equipment.

Figure 6-1-1. This shows a single pole circuit breaker (left) that would be used on a 120 VAC power source and a 2-pole circuit breaker (right) that would be used for protecting a 208/230 VAC system.

PROCEDURE

Step 1. Your instructor will assign you five HVACR units with readable nameplates. Record unit model number or first 8 digits, unit voltage, minimum circuit ampacity, maximum circuit ampacity, wire size that should be used to install this unit, and the breaker or fuse size to be installed with unit.

Figure 6-1-2 is an example of a unit nameplate. Notice the compressor and refrigerant information at the top, fan motor information in middle of nameplate, and conductor sizing information under the fan motor information. The first four digits of a Carrier serial number indicate the date the unit was manufactured. The first and second digits indicate the week and the third and fourth digit indicate the year it was made. The serial number of the nameplate tells us that this unit was made the 37th week of 2011.

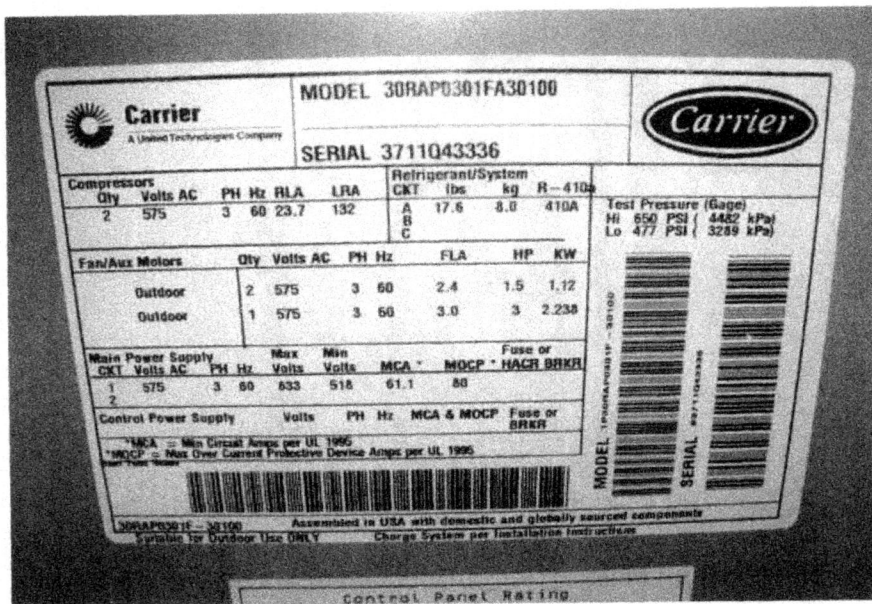

Figure 6-1-2. This is a sample nameplate from a Carrier unit.

Use Table 6.4 from the text, NEC table 310.15, to determine the fuse or breaker size. Record your data below.

Unit M/N	Voltage	Minimum Circuit Ampacity	Maximum Circuit Ampacity	Wire size	Fuse or breaker size

Step 2. Acquire four sections of wire with insulation on the wire. The insulation should have the size and insulation type stamped on it, as shown in Figure 6-1-3.

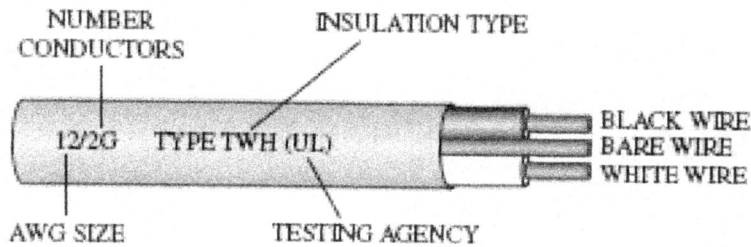

Figure 6-1-3. This is an example of the information on Romex wire. The gauge is 12 AWG, with two conductors. The *G* indicates a bare ground wire. The insulation type is TWH and it is UL approved.

Use Table 6.4 from the text, NEC table 310.15, to determine fuse or breaker size. Record your data below.

Wire size	Wire type (copper, Alum.) etc.	Insulation type	Temp. rated	Amperage rating of wire

Step 3. Select five electric motors in your lab and complete the table below. From the motor nameplate, record the name brand or type of motor, voltage or voltages of motor if dual voltage, phase (1 or 3), RLA or FLA, motor LRA, motor horsepower, and frequency or hertz. Refer to Section 6.7 in the text for information on common motor nameplates per NEC.

Name brand of motor	Voltage	Phase	Motor RLA or FLA	Motor LRA	Motor horse power (HP)	Frequency or hertz

QUESTIONS

1. Wire is sized using AWG table, what does AWG stand for?

2. Does 4 AWG aluminum wire carry the same amperage as 4 AWG copper wire?

3. Which size wire is the largest: 10 AWG or 14 AWG?

4. A 10 AWG THWN wire has an ampacity of?

True or False

_____ 5. A motor fuse size is determined by the LRA.

_____ 6. The breaker size for a unit is determined by the smallest circuit amps.

_____ 7. NEC allows fuse size of up to 225% of a compressor's RLA.

_____ 8. Orange colored Romex wire only comes in size 10 AWG.

Unit 7 Electrical Installation of HVACR

Unit Summary

Installing power and control voltage wiring is not the major time-consuming element when it comes to installation or equipment change-out—but it is the most important step in the installation because improper wiring may create a hazardous condition that can cause a fire or electric shock hazard to the technician or the consumer. The wire size, overcurrent protection, and proper grounding are the important safety aspects of good installation or system exchange. When doing calculations for selecting wiring and overcurrent protection, it is a safe practice to round down for overcurrent protection and round up for wire size.

When troubleshooting an electrical problem that involves "butchered" wiring, it may be easier to rewire it rather than spend time trying to figure out the problem. After removing the wiring, start by hooking up the control voltage circuit first. Test each circuit as it is wired. Next, hook up and test components in the high-voltage section. Start with the easiest component first. For example, if you are rewiring a condensing unit, test the control voltage circuit first. In the high-voltage section, wire the crankcase heater or condenser fan motor first. Proceed with more complex components like the compressor.

Key Terms (Definitions can be found in the Glossary in your text.)

Annual fuel utilization efficiency

Bonding jumper

Control voltage

Crankcase heater

Ferrule

Green installation

Hard start kit

Liquid line solenoid valve

Maximum circuit ampacity

Maximum overcurrent Protection

Open wiring

Plenum-rated cable

Seasonal energy efficiency ratio

Stranded wire

LAB 7.1 HVACR Electrical Installation and Checklists

LABORATORY OBJECTIVE

You will use the data plate of a unit to correctly size the wire, disconnect and breaker switch for the unit, perform a prestart electrical check, and demonstrate how the thermostat operates.

ELECTRICITY FOR HVACR, 1e TEXT REFERENCE

Unit 7: Electrical Installation of HVACR

REQUIRED MATERIALS PROVIDED BY THE STUDENT

- Bulleted list of materials and equipment

- 6-in-1 screw driver

- Multimeter

- Wire cutters and strippers

REQUIRED MATERIALS PROVIDED BY THE SCHOOL

- Working package unit with programmable thermostat installed, but line voltage and disconnect unwired

- Wire

- Conduit and connectors

- Disconnect switch

SAFETY REQUIREMENTS

Follow all lab and shop safety rules, which includes wearing safety glasses at all times. Adhere to lockout/tagout procedures and never assume power is off, always check with a multimeter. It is unsafe to work around live electricity while wearing a watch or rings, so remove them before starting this task. Units and disconnect can have sharp edges so comfortable gloves are recommended.

INTRODUCTION

In this lab assignment you will be following the general procedures for installing a split system air conditioner. As an installing technician, you never want to install a system without checking and verifying proper operation according to the manufacturer. It is good practice to check certain conditions of the system so that the equipment owner can be assured that their equipment is operating as it was designed to and they are getting the most for their money. Post installation checks also help to reduce callbacks, which can be costly to a contractor and negatively affect their reputation.

Manufacturers often provide checklists and charts to help you achieve maximum efficiency and reliability of the equipment; remember the manufacturer's reputation is on the line as well so it benefits them to make sure these checks get done properly. You should use the checklists after every installation. Electrical items that are often included with air conditioners are amperage checks, voltage checks, thermostat programming checks, and general component operation checks (i.e., make sure that the compressor runs and sounds normal, make sure the condenser and evaporator fans are both operating and make sure the thermostat controls the system correctly). You should find the applied voltage within +/- 10% of the voltage listed on the nameplate and current draw should be at or near the full load value (FLA or RLA) when the system is under a load. Manufacturers will also usually suggest that you check the tightness of electrical connections and proper wiring according to the wiring diagram that is supplied with the unit and in the installation guide. There are usually some mechanical system checks that need to be performed, but since this is an electrical lab we will focus only on electrical check items. It is also a good idea to inform the homeowner or building manager how the thermostat operates and is programmed, how the system should sound and perform, and how to do some basic maintenance checks in addition to telling them how often this maintenance should be performed. Lastly, you should give the installation guide to the homeowner and tell them to keep it in a safe place where they can find it; don't throw it away and don't take it with you unless you plan on filing it in a safe place at your company shop. They usually contain important service and troubleshooting information that can be referenced later.

PROCEDURE

Step 1. See your instructor for an assigned air conditioning unit that is in working order with control wiring installed.

Step 2. On your assigned unit data plate find the minimum circuit ampacity to size power wire and disconnect and the maximum overcurrent protection on the unit to size the fuse or breaker switch for the unit. Figure 7-1-1 is an example of a data plate.

CONTAINS HCFC – 22		DESIGN PRESSURE		
FACTORY CHARGE		278		HI PSIG
12 LBS	8 OZS	144		LO PSIG
ELECTRICAL RATING		NOMINAL 208/230		VOLTS
1 PH	60 HZ	MIN 197	MAX 253	
COMPRESSOR(S):(1)		**FAN MOTOR(S): (1)**		
PH	1	PH	1	
RLA	23.8	FLA	1.7	
LRA	129	HP	1/4	
MIN. CKT AMPACITY AMPERAGE MINIMUM	31.5	MAX FUSE OR CKT.BKR. FUSIBLE/COUPE CIRCUIT (HACR PER NEC)	50	
FOR OUTDOOR USE				
VERIFIED		VERIFIE		

Figure 7-1-1. This is an example data plate that contains important data on electrical specifications, refrigerant type and charge, and individual components.

The disconnect should be rated for no less than 115% of the minimum circuit amps. Refer to Table 6.4 in Unit 6 for wire size.

Unit model number:	Disconnect ampacity:
Minimum and maximum voltage:	Wire size:
Minimum circuit ampacity:	Breaker or fuse size:
Maximum overcurrent protection:	

Step 3. Mount the disconnect if not mounted already. See Figure 7-1-2, which shows two types of disconnects. On the left is a pull disconnect for a condensing unit rated for 60 A, which will more than handle the total load of the condenser, which is 25 A. The "pull" is sitting on the top of the disconnect, ensuring that no power is supplied to the condensing unit. On the right: The blades on a throw disconnect do not always open or close. A visual check of the disconnect is one way to be certain it is open and safe.

Wire the unit and disconnect per installation instructions. ***Do not turn power on!*** Have your instructor check your work for safe, tight, and correct connections.

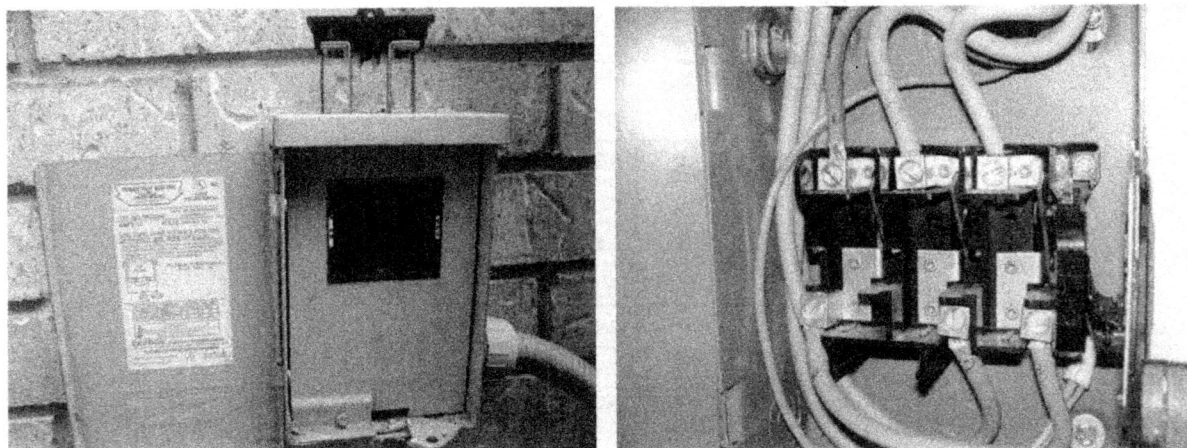

Figure 7-1-2. This figure shows two types of disconnects.

Step 4. With disconnect off, turn power on and check line voltage at disconnect as shown in Figure 7-1-3.

Line voltage at disconnect with disconnect Off _____

Figure 7-1-3. This shows a technician checking line voltage at disconnect. Notice the technician is not wearing a watch or rings, which can conduct electricity, however holding the meter and holding probe in each hand is an unsafe practice.

Step 5. If voltage is within allowable range turn disconnect on and check voltage at unit.

Line voltage at unit with disconnect on, unit Off _____

Step 6. Turn unit on and check incoming voltage at unit, it should still be within allowable range. If there is a difference in the voltages with unit off and unit on, this is the voltage drop.

Line voltage at unit, unit on and running _____

Voltage drop   _____

Step 7. Check and record total amperage of unit. Total amperage should be close to minimum circuit ampacity.

Total Amperage _____

Step 8. Thermostat should be mounted at 60 inches above finished floor (5 feet high).

Show your instructor you understand how to operate the thermostat by completing the task below.

Current temperature _____

Turn fan on.

Fan auto with call for cooling.

Fan auto with call for heat.

Emergency heat on if Heat Pump.

Program thermostat to the schedule given to you by your instructor.

Explain how to operate thermostat as if you are explaining it to a new customer.

QUESTIONS:

1. What color is the ground wire?

2. How high above the floor is a residential T-Stat mounted?

3. Where should a disconnect switch be mounted in reference to the unit?

4. What is the allowable voltage range for a 208 – 240 volt HVACR system?

5. The power that feeds from an electrical panel to a disconnect switch is connected to the

   _____ terminal in the disconnect.

LAB 7.2 Introduction to Diagrams

LABORATORY OBJECTIVE

To get you familiar with wiring diagrams and locating the electrical components list in the diagrams in an actual heat pump unit.

ELECTRICITY FOR HVACR, 1e TEXT REFERENCE

Unit 7: Electrical Installation of HVACR

Unit 8: Transformers

Unit 9: Relays, Contactors, and Motor Starters

Unit 10: Capacitors

REQUIRED MATERIALS PROVIDED BY THE STUDENT

- 6-in-1 screwdriver

- Multimeter

REQUIRED MATERIALS PROVIDED BY THE SCHOOL

- Heat pump split system, system does not have to be in working order

SAFETY REQUIREMENTS

This lab is not to be done with power off, so check incoming power with multimeter. Be sure to follow lockout/tagout procedures for the breaker or disconnect if the system is in working order. Care should be taken when removing unit cover to prevent injuring your hands.

INTRODUCTION

Unit wiring diagrams are important to installation and service technicians alike. They show the installer the location of incoming power connection and many times the voltage requirements. They also show the location of low voltage connections and the corresponding indoor thermostat connections. The diagram shows the service technicians the wire colors, which is useful when troubleshooting. It identifies the parts that can be available for installation on the unit if the service tech feels it is necessary. Like a hard-start kit for systems with a long refrigerant line set or a crankcase heater for systems installed in low ambient temperature. From the wiring diagram a service technician can determine the normal operating procedure, which is a must for troubleshooting problems. The first step in understanding these diagrams is a visual inspection to locate the actual unit components listed in the diagram. Take a close look at the diagram to determine this basic information.

PROCEDURE

Step 1. Remove the covers from the heat pump split system you have been assigned. Examine the outdoor unit and locate the items listed below.

Outdoor unit

- Condenser fan motor
- Compressor
- Contactor
- Capacitor
- Reversing valve
- Indoor fan motor

Examine the indoor unit and locate the items listed below.

Indoor Unit

- Indoor fan motor

- Transformer

- Electric heating elements

- Heater relay

- Transformer

- Fan relay

Be prepared to identify the items upon instructor request. Replace unit covers when you have completed task.

Step 2. Carefully examine the diagram in Figure 7-2-1 and answer the questions below. To see this diagram in color, refer to Figure 7-4 in your textbook.

START RELAY

BLACK

⑤ ②

①

HARD START,
IF USED

YELLOW

START
CAPACITOR

RED

OUTDOOR
FAN

PURPLE
BLACK
ORANGE
RED
YELLOW

DUAL
CAPACITOR

Ⓕ Ⓒ Ⓗ

RED

RED

A4
TIMED OFF
CONTROL, IF USED

② ③ ①

TO 24 VAC POWER
SOURCE
20 VA MINIMUM
NEC CLASS 2

C   Y1

Ⓡ Ⓢ Ⓒ

COMPRESSOR

BLACK

COMPRESSOR
CONTACTOR

CRANKCASE HEATER, IF USED

⚠

L2
L1
208-230/60/1
GROUND

GROUND
LUG

⚠ L1

208-230/60/1

L2

EQUIPMENT
GROUND

K1 - 1

BLACK

HR 1

BLACK

⑤ R1

①

② ①

YELLOW

Ⓗ

C 12

YELLOW

RED

C 23

RED

BLACK

RED

Ⓒ Ⓢ Ⓡ

B1

Ⓒ Ⓕ

B4

PURPLE

ORANGE

S4
HIGH
PRESSURE
SWITCH, IF USED

S24
LOSS OF
CHARGE
SWITCH, IF USED

S4

K1

② A4
③
①

S24

C   Y1

TO 24 VAC
POWER SOURCE
20 VA MINIMUM
NEC CLASS 2

KEY	COMPONENT
A4	CONTROL-TIMED OFF
B1	COMPRESSOR
B4	MOTOR-OUTDOOR FAN
C12	CAPACITOR-DUAL
C23	CAPACITOR-START
HR1	HEATER-COMPRESSOR
KL-1	CONTACTOR-COMPRESSOR
R1	RELAY-START
S4	SWITCH-HIGH PRESSURE
S24	SWITCH-LOSS OF CHARGE

← INDICATES OPTIONAL
COMPONENTS

⚠ FOR USE WITH COPPER CONDUCTORS ONLY.
REFER TO UNIT RATING PLATE FOR MINIMUM
CIRCUIT AMPACITY AND MAXIMUM OVERCUR-
RENT PROTECTION SIZE.

WARNING-
ELECTRIC SHOCK HAZARD. CAN CAUSE INJURY OR DEATH.
UNIT MUST BE GROUNDED IN ACCORDANCE WITH
NATIONAL AND LOCAL CODES.

————————— LINE VOLTAGE FIELD INSTALLED
— — — — — CLASS II VOLTAGE FIELD INSTALLED
▬ ▬ ▬ ▬ ▬ THERMOSTAT WIRING

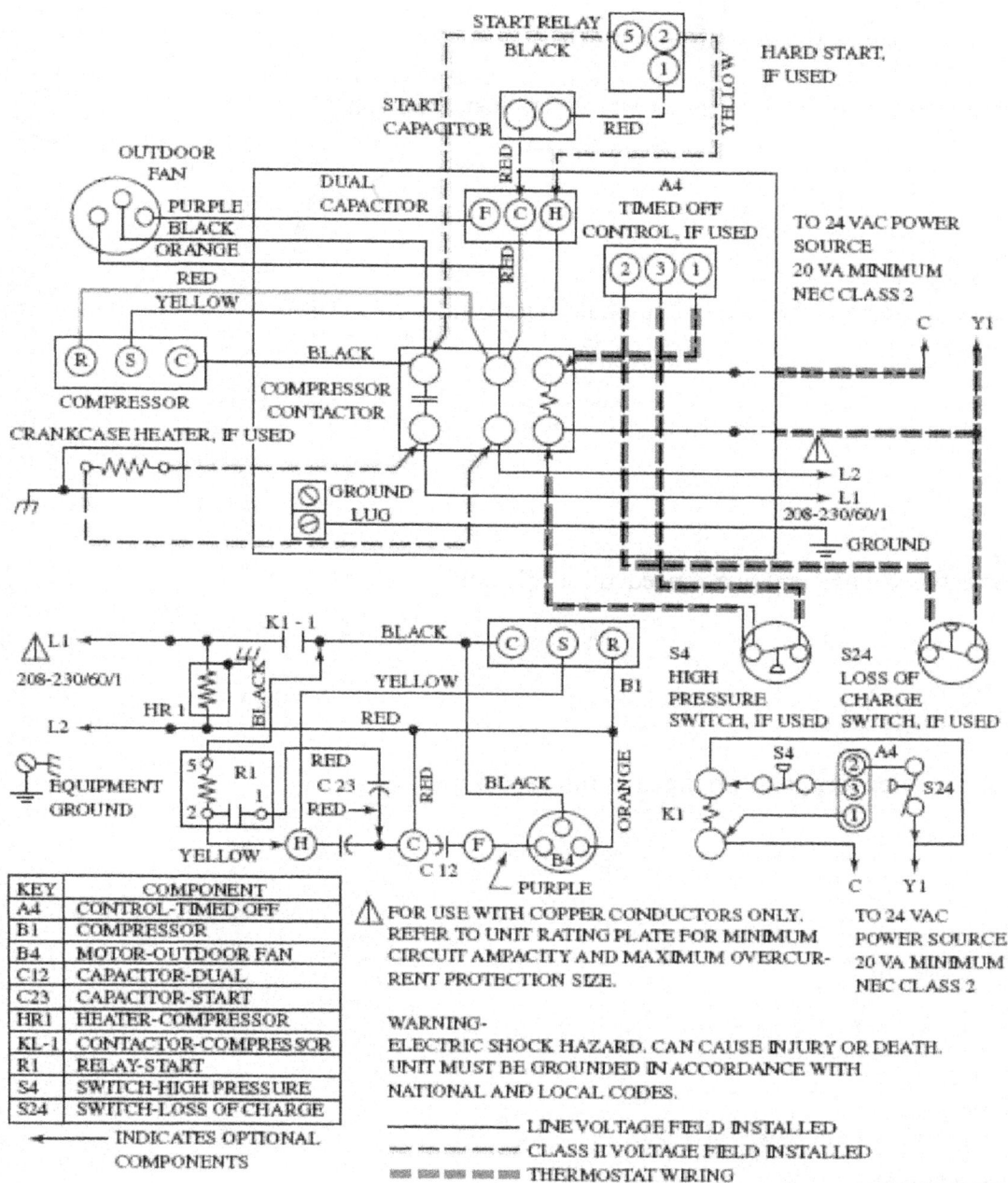

100458-01, REV-F

Figure 7-2-1. An installation diagram. Notice in this diagram that the voltage source is 208/230 and 24 V. The dashed lines represent field-installed wiring that may be low voltage or high voltage. The highlighted yellow dashed lines are high-voltage, field-wired circuit options. The red dashed lines are low-voltage thermostat wiring.

QUESTIONS

Use Figure 7-2-1 to answer all questions below

1. What terminals on the thermostat are used to install this unit?

2. What component is the line voltage field wire connected to?

3. What is the voltage and phase needed for this unit?

4. Name the three terminals on the dual run capacitor.

5. What is HR 1?

6. What color wires go the outdoor fan?

7. What is minimum VA for the low voltage circuit?

Unit 8 Transformers

Unit Summary

It is important to understand how transformers operate because they are one of the first pieces of equipment that should be checked when troubleshooting. If the input voltage is present, you can determine if that part or all of the system has voltage. If the output voltage is present, you can continue the troubleshooting process on the control voltage side. This troubleshooting process will depend on the problem and may not be the most expedient way to proceed. It is a place to start if you do not know anything about the circuit operation.

The first step to checking a transformer is determining if there is input voltage. Next check the output voltage. Some transformers burn out due to unknown reasons. Some burn out due to overloading (too much current draw) on the secondary side. Others burn out due to shorted components in the secondary side of the electrical circuit. Check for circuit overloading by measuring the amp draw on the secondary side of the transformer. Check for short circuits or low-resistance circuits with an ohmmeter prior to attaching the secondary of the transformer. Place a properly sized fuse in the secondary of a replacement transformer. This can be a temporary fuse that you can remove and wire directly or, better yet, a permanent fuse that will protect the transformer from a short-circuit condition. Size the fuse at 150% of the secondary amperage rating to prevent nuisance blown fuses, yet still protect against a short-circuit condition. This will not protect against an overload condition. An overload condition is one in which a higher than normal amp draw is occurring, which will overheat the transformer and eventually burn it out. Use the touch test. A transformer will be slightly warm when fully loaded. It should never be hot to the touch. Hot transformers will soon burn out.

Key Terms (Definitions can be found in the Glossary in your text.)

Commercial transformer

Industrial transformer

Multi-tap transformer

Transformer

Volt-amps

LAB 8.1 Transformer ID and Analysis

LABORATORY OBJECTIVE

You will identify transformers by the input VAC, output VAC, and size by VA. You will then check the condition of transformers by performing a resistance check. Finally, you will change a transformer in a working unit.

ELECTRICITY FOR HVACR, 1e TEXT REFERENCE

Unit 3: Safe Use of Electrical Instruments

Unit 8: Transformers

REQUIRED MATERIALS PROVIDED BY THE STUDENT

- 6-in-1 screwdriver

- Needle nose pliers

- DMM or VOM suitable for HVACR field work

REQUIRED MATERIALS PROVIDED BY THE SCHOOL

- Six transformers of various primary and secondary voltages

- Working air conditioning unit to replace transformer

SAFETY REQUIREMENTS

Make sure power is off and disconnect is locked out when changing parts on an air conditioning system. Since you will be working on the unit with power on and checking voltage, care should be taken not to touch exposed connection. Always use needle nose pliers to remove spade connectors to protect your hands. Always use common sense and follow all lab safety rules. As with any laboratory activity, safety glasses should be worn at all times.

INTRODUCTION

Transformers are the component responsible for control voltage in an air conditioning system and are one of the first checks a service technician makes on a service call. It is important to know how transformers work as well as how to determine if a transformer is good or bad. The correct transformer for a unit is determined by several factors. The incoming voltage or primary volts are the voltage supplied to the transformer and usually the same as voltage supplied to the equipment. Many transformers have multiple taps so the transformer can be used in different applications and different supply volts. Figure 8-1-1 shows a transformer with multiple input voltage options. The outgoing or secondary VAC is the voltage coming out of the transformer. Secondary voltage is supplied to the thermostat and is responsible for energizing control components. The size of a transformer is listed in VA, which is calculated by volts × amps.

Troubleshooting transformers should be done in two steps with the first being voltage checks. Check voltage on the primary and then secondary windings. If you have voltage on primary and none on secondary, the transformer is bad. Next is to determine how it is bad by resistance checks on primary winding and then on secondary winding. This will determine if the transformer has an open or shorted winding.

Figure 8-1-1. A transformer with multiple input voltage options, which is helpful to a service technician because one transformer can be used to replace several different applications.

Figure 8-1-2. A label showing how to wire a transformer with multiple voltage options. With this transformer the common wire is yellow and should always be connected, if unit's supply voltage is 200 VAC connect yellow and red, if unit's supply voltage is 460 connect yellow and black. Always remember to cap the transformer leads not being used.

PROCEDURE

Step 1. Obtain six transformers from your instructor that are not installed in a working unit.

Record the types of transformers (step up or step down). Record the primary and secondary voltage of transformers. Check and record the resistance of the primary and secondary windings with a VOM. If the transformer is a multi tap, measure the resistance between common and each tap separately. Care should be taken not to touch the meter leads with your fingers when checking resistance. Record the size and the condition of the transformer (whether it is good, open primary or secondary winding; shorted primary or secondary winding).

Type	Primary voltage	Primary resistance	Secondary Voltage	Secondary resistance	Size in VA	Condition

Step 2. Change a transformer in an air conditioning unit by following the steps below. Your instructor will assign you an air conditioning unit to replace the control transformer. Determine the type of transformer needed for replacement, verify the primary and secondary voltage, and check the resistance of replacement transformer to make sure it is good.

Step 3. Disconnect unit from power source, lockout the disconnect.

Step 4. After writing down the old transformer wiring, remove from the unit.

Step 5. Install new transformer wired like the old transformer. Have instructor check your installation.

Step 6. Turn supply power on and check primary and secondary voltage.

Record supply voltage _____

Secondary voltage _____

Step 7. Turn thermostat on a call for cooling and check the amperage of transformer.

Transformer amperage_____

Size fuse protecting transformer_____

QUESTIONS

1. When checking a transformer the primary voltage is 240 and the secondary voltage is zero. Does this mean the secondary windings of the transformer are bad? Explain.

2. When checking amperage on a low voltage transformer, the meter reads 35 amps and has 10 wraps around amp meter. What is the actual amperage of the low voltage circuit?

3. The resistance of a secondary winding of a transformer is 0.000. What is the condition of the transformer?

4. The resistance reading of the primary winding of a transformer is OL. What is the condition of transformer?

5. What size low voltage fuse should be used in a unit with a 60 VA 24 volt secondary transformer?

6. What size low voltage fuse should be used in a unit with a 60 VA 120 volt secondary transformer?

7. The two connections on the secondary side of a low voltage transformer are labeled as what?

Unit 9 Relays, Contactors, and Motor Starters

Unit Summary

This unit covered the selection, operation, and troubleshooting of relays, contactors, and motor starters. These electromechanical devices are the switches used on HVACR fans, pumps, and compressors. It is important to know the fundamental operation of each device in this family of switches. Common selection criteria include knowing the coil voltage, contact amperage, voltage rating, and number of contacts required for a replacement unit.

This unit included troubleshooting tips. Many loads, such as motors, are replaced unnecessarily because the technician does not understand the controlling switches such as the relay, contactor, or motor starter. Understanding these components leads to a big advancement in a technician's ability to troubleshoot HVACR systems.

Key Terms (Definitions can be found in the Glossary in your text.)

Amps full load (AFL)	Motor starter
Amps locked rotor (ALR)	Overload protection
	Relay
Auxiliary contacts	Single-pole contactor
Chatter	Start
Contactor	Stop
Contacts	Three-pole contactor
De-energized state	Two-pole contactor
Four-pole contactor	

LAB 9.1 Relay, Contactor, and Motor Starter
Identification and Specifications

LABORATORY OBJECTIVE

In this lab you will learn how to identify relay, contactor, and motor starter specifications and select them for various HVACR applications.

ELECTRICITY FOR HVACR, 1e TEXT REFERENCE

Unit 9: Relays, Contactors, and Motor Starters

REQUIRED MATERIALS PROVIDED BY THE STUDENT

- DMM or VOM suitable for HVACR field work

REQUIRED MATERIALS PROVIDED BY THE SCHOOL

- Three numbered relays

- Three numbered contactors

- Three numbered motor starters

- Sales catalog that lists contactors

SAFETY REQUIREMENTS

Since we are not dealing with any "live" electricity for this laboratory activity there is no risk of electric shock. Some common sense safety precautions do apply though. As with any laboratory activity, safety glasses should be worn at all times.

INTRODUCTION

Relays, contactors, and motor starters are all variations on the relay concept. All of these devices contain two basic parts; a coil and one or more sets of contacts. The coil operates the contacts when the correct voltage is applied to them through electromagnetism. It is critical that the coil component (which is a load) has the correct voltage applied to it. All relays, contactors, and motor starters will show the correct coil voltage on the label. If this label is worn off or missing, you can usually determine the correct voltage by examining the circuit it is connected to or the unit wiring diagram, if available.

The contacts operate one or more loads in another circuit. Since these contacts are a type of switch, they are rated at the maximum current they can handle at a given voltage. It is important that you do not exceed the contacts' amperage rating otherwise relay failure or worse, failure of the load it controls, will result. Several values may be listed to help you select a relay, contactor, or motor starter with the right size contacts. AFL (amps full load) or FLA (full load amps) is usually the number of interest; AFL and FLA mean the same thing. Others you may see are ALR (amps locked rotor) and LRA (locked rotor amps), or a rating in motor horsepower (HP). Again, for clarity, ALR and LRA mean the same thing.

The contact arrangement can be of many different types. Relays commonly come in SPST, SPDT, and DPDT types. Contactors and motor starters, which generally control larger loads than relays, are almost always single throw and come in SPST, DPST, or triple pole single throw (TPST) arrangements. More than three sets of contacts (poles) can be found on some. The only exception is if a contactor or motor starter is fitted with an auxiliary contact accessory, which may close when the main contacts close, or open when the main contacts close.

All of these contacts are given the designation normally open (NO) or normally closed (NC) to describe the position the contacts are in when there is no voltage present at the coil. Contacts are always shown in their normal position in a wiring diagram. Figure 9-1-1 shows some common diagram symbols used for relay, contactor, or motor starter components. Other symbols are sometimes used, especially for coils. You should review Unit 9 in your textbook for other symbols and contact variations.

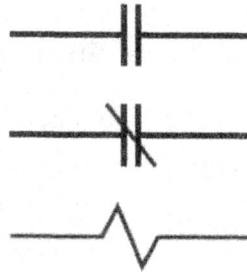

Figure 9-1-1. This shows common diagram symbols used for relay parts. From top: normally open contacts, normally closed contacts, coil.

PROCEDURE

Step 1. Using the three numbered relays, fill in the table identifying characteristics:

Relay #	Coil voltage	Number or normally open contacts	Number of normally closed contacts	Contact switch action (SPST, SPDT, etc.)	FLA or AFL Contact amperage rating (10A @ 120VAC, etc.)
1					
2					
3					

Step 2. Using the same three relays, locate the two terminals where the coil is located for each relay. Using your VOM check the resistance of each coil and record them below. Are any of these coils faulty?

Step 3. Using the three numbered contactors, fill in the table identifying characteristics:

Contactor #	Coil voltage	Number or normally open contacts	Number of normally closed contacts	Contact switch action (SPST, SPDT, etc.)	FLA or AFL Contact amperage rating (10A @ 120VAC, etc.)
1					
2					
3					

Step 4. Using the same three contactors from step 3, locate the two terminals where the coil is located for each contactor. Using your VOM check the resistance of each coil and record them below. Are any of these coils faulty?

Step 5. With the motor starter you were supplied, fill in the table identifying characteristics:

Motor Starter #	Coil voltage	Number or normally open contacts	Number of normally closed contacts	Contact switch action (SPST, SPDT, etc.)	FLA or AFL Contact amperage rating (10A @ 120VAC, etc.)
1					
2					
3					

Step 6. Using the same three motor starters from step 5, locate the two terminals where the coil is located for each starter. Using your VOM check the resistance of each coil and record them below. Are any of these coils faulty?

QUESTIONS

1. Looking back at your answers from the steps, what is the major difference between the relays and the contactors that you examined?

2. What is different about the motor starters compared to the other devices?

3. The motor starters have at least one reset button on them. Why does a motor starter have a reset button, but the contactors and relays do not?

4. Some contactors and motor starters can be repaired when they fail. What parts of your contactors and motor starters can be replaced?

5. Relay, contactor, and motor starter contacts are always shown in their _____ position in a wiring diagram.

To answer questions 6-9, refer to ED-14, which appears with the Electrical Diagrams package that accompanies the text.

6. Using ED-14, identify all of the relays, contactors and motor starters in this system.

7. For the device labeled "R1" in the diagram, how many contacts does it have? Are they normally open or normally closed?

8. There are two devices labeled "P1" and "P2" in the diagram. Where are the coils located in the diagram for these devices? What switches are in series with these coils?

9. The coils for "P1" and "P2" from question #8 are in series with "P1-OL" and "P2-OL" respectively. What do you think is the purpose of having switches from the same motor starter in series with their own coils?

10. You need to replace a relay that powers an indoor blower motor in an air handler. The motor is a ½ HP motor that operates at 120 VAC. The AFL of the motor is 6.5 amps. The relay coil requires a 24 VAC signal from the thermostat to operate. Using the catalog supplied by your instructor select a relay for this blower motor.

11. You need to replace a contactor on an air conditioning condensing unit. The unit has the following information printed on the nameplate: FLA 22.6, LRA 135, 208-230 VAC. The system has a transformer and operates with a 24 VAC control circuit. Using the catalog supplied by your instructor select a contactor for this condensing unit.

LAB 9.2 Relay Logic – Part 1

LABORATORY OBJECTIVE

In this lab you will learn how to wire general-purpose relay diagrams and understand the

function of any relay, contactor, or motor starter in an electrical circuit.

***ELECTRICITY FOR HVACR, 1e* TEXT REFERENCE**

Unit 9: Relays, Contactors, and Motor Starters

REQUIRED MATERIALS PROVIDED BY THE STUDENT

- DMM or VOM

REQUIRED MATERIALS PROVIDED BY THE SCHOOL

- A basic wiring trainer with several relays

SAFETY REQUIREMENTS

This lab will involve voltage measurements on "live" circuits. Do not plug in the circuit until you

have received the approval of your instructor. Remember to follow safe hand tool use practices

and make all electrical connections neat to prevent wires from accidentally touching. This lab

will involve more complicated wiring so you need to take extra precautions to prevent wires

from touching each other or ground. Lockout/tagout procedures should also be followed. When possible you should turn off the power supply before connecting the meter to take a reading, then you can turn the power back on. You should know how to quickly turn power off to the circuit in the event of a problem, which is usually by pulling out a plug or turning off a main switch.

INTRODUCTION

As we learned in Lab 9.1, relays, contactors, and motor starters control the function of one circuit with another, separate circuit. For this lab we will be examining a couple of circuits, drawing a wiring diagram for each, and then proceed to wire each diagram on a trainer using your hand-drawn diagram.

The ability to develop a wiring diagram for a system, no matter how basic or complex, is a challenging task but one that technicians are required to do frequently. Wiring diagrams that are provided by manufacturers are not always complete and none of them are the same. Sometimes you will not have a diagram available or the diagram was destroyed or not legible. Other types of systems are custom built and you must make a diagram for it. Labs 7.2 and 17.1 provide you with the building blocks of wiring diagrams, now we will use this information to develop more complicated diagrams with relays, contactors, and motor starters, which are found in nearly all HVACR systems.

PROCEDURE

Step 1. Reading the following description of a circuit and then develop a ladder-type diagram for this circuit using the symbols and rules learned from previous labs. The loads in the diagram will be shown simply as a circle with the load letter inside. For example, "Load A" will be shown as the symbol in Figure 9-2-1.

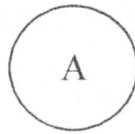

Figure 9-2-1. This is the symbol to use for load "A" in your diagrams. For other loads simply insert the appropriate letter in a circle.

A circuit contains two loads that are wired in parallel with each other, loads "A" and "B." Both loads are switched by one relay (R1) in a 120 VAC circuit. When "A" is off, "B" will be on, so the relay contacts need to have a SPDT arrangement. The coil of the relay is in a 24 VAC control circuit and is operated by one SPST switch. Draw your ladder diagram for this system in the space below.

Step 2. Get approval from your instructor for this diagram and then proceed to wire it into your trainer. Before applying power to the circuit get the approval of your instructor.

Step 3. Read the following description of a circuit and then develop a ladder-type diagram for this circuit in the space provided.

This circuit has four 115 VAC loads (Loads A, B, C, and D). A DPDT relay contact arrangement will be used to operate these loads. Loads A and B should be energized when C and D are de-energized and vise versa. The coil of this relay is operated with 24 VAC and is in series with two SPST switches in the control circuit.

Step 4. Get approval from your instructor for this diagram and then proceed to wire it into your trainer. Before applying power to the circuit get the approval of your instructor.

Step 5. Read the following description of a circuit and then develop a ladder-type diagram for this circuit in the space provided.

This circuit has three relays. Relay 1 is SPST and controlling load "A." Relay 2 is SPDT and is controlling loads "B" and "C." Relay 3 is an SPST contactor that is controlling all three loads. When the coil of relay 1 is energized by a SPST the switch in the control circuit load "A" will turn off. For relay 2, when the coil is energized, load "B" will turn off and "C" will turn on, the coil of relay 2 is operated by another SPST switch. Relay 3 (the contactor) will be normally open and its coil will be operated by a third SPST switch. All coils are operated with 24 VAC.

Step 6. Get approval from your instructor for this diagram and then proceed to wire it into your trainer. Before applying power to the circuit get the approval of your instructor.

QUESTIONS

Completion of this lab is based only on your demonstration of techniques. There are no associated questions to answer.

LAB 9.3 Relay Logic – Part 2

LABORATORY OBJECTIVE

For this lab you will be working with more advanced diagrams that contain relay circuits. Many HVACR systems have circuitry that contains multiple relays and their function and operation are important to understand.

ELECTRICITY FOR HVACR, 1e TEXT REFERENCE

Unit 9: Relays, Contactors, and Motor Starters

REQUIRED MATERIALS PROVIDED BY THE STUDENT

None

REQUIRED MATERIALS PROVIDED BY THE SCHOOL

None

SAFETY REQUIREMENTS

None

INTRODUCTION

This lab activity will utilize the ED-01 diagram packaged with your text.

Relays control the functions of many loads within an HVACR system. Motors, heater elements, valves, and other loads often are controlled by a relay contact instead of simply using a switch. A simple switch without a relay could control all of these loads, but there are certain advantages to using relay-operated circuits that we will look at in this lab.

Firstly, larger loads generally pull more current. Therefore, the switch that controls the larger loads must be built for handling higher current. Many of the switches we use are sensitive pressure or temperature sensing devices, which would be difficult to design for higher current without sacrificing some sensitivity. Another common problem is that many HVACR systems are remotely controlled. We discussed thermostats for split type furnaces and air conditioning systems in Lab 7.1. If these thermostats were to operate the air conditioning system directly without a relay they would need to be line voltage and be built to withstand a lot more. This is a safety concern as well as an expense concern. Instead, a 24 VAC control circuit with one or more relays are used, which is much safer to operate and reduces the current needed, which in turn allows installers to use much smaller (and much cheaper) thermostat wire.

PROCEDURE

Step 1. Using ED-01, identify the number of relays in the entire circuit and record the name of each one below.

Step 2. Identify the location of the 24 VAC low voltage control circuit and the 120 VAC line voltage circuit in the diagram.

Step 3. Using the list you made in Step 1, identify the number of contacts that each relay contains. Also identify if each contact is normally open or normally closed.

Step 4. Using the information you collected in all of the previous steps, identify the load that each relay controls. For example, the compressor relay (CR) operates the compressor motor (C). You do not need to include other switches, such as pressure controls or thermostats.

QUESTIONS

1. Why do you think the relay labeled HR has two sets of contacts (a NO and NC) that both operate the IFM relay coil?

2. What type of heat is this system using?

3. The thermostat in this system does not use a common wire (C) going to the transformer. What type of thermostat does?

4. If you were to measure the current of all the loads in the line voltage circuit and add them up, do you think it would be higher or lower than the current in the control circuit? What advantages does this provide?

5. There are several safety controls used in this diagram, list them below. One thing you may notice is that none of these safeties involve the use of a relay. Why do you think this might be?

LAB 9.4 Troubleshooting Control Circuits

LABORATORY OBJECTIVE

In this lab you will learn how to use a DMM or VOM to troubleshoot common control circuits.

ELECTRICITY FOR HVACR, 1e TEXT REFERENCE

Unit 9: Relays, Contactors, and Motor Starters

REQUIRED MATERIALS PROVIDED BY THE STUDENT

- DMM or VOM suitable for HVACR field work

REQUIRED MATERIALS PROVIDED BY THE SCHOOL

- An HVACR system or wiring trainer with a control circuit problem

SAFETY REQUIREMENTS

For this lab you will be working with a piece of equipment with power applied so safety

precautions should be taken. Remember to follow safe hand-tool use practices. Do not power up

the equipment until your instructor gives you the ok.

INTRODUCTION

There are control circuits found in most HVACR systems. Some of them operate with 24 VAC

and some operate at higher voltages. With either type you need to know how to effectively

troubleshoot a control circuit to find problems with the system. Typically, control circuit

problems will be presented as a major component failing to run. Before beginning the

troubleshooting process, the first thing you should do is make sure the system is running by

checking the thermostat and then identifying which loads are operating and which are not. If any load that should be operating is not when there is a call for heat or cooling, the problem is electrical and you should locate the circuit where the problem load(s) are located.

Take for example ED-04 found packaged with your text. At first glance, electrical troubleshooting this system looks difficult at best. But you really only have to look at a small piece of the circuit. Let's assume you find that the compressor is the load not starting. The first step is to locate the compressor in the diagram, which is the bottom-most line voltage load. This is the only circuit you need to focus on at the moment. You should verify that voltage is available to the compressor starter (CS) at the far left and far right of the branch where the compressor is located. If it is not, there is a problem with the power source and you need to check there. If voltage is present, you will begin the troubleshooting process known as "hopscotching."

Hopscotching is a very powerful tool to use when finding problems in electrical circuits. It involves taking voltage measurements within a circuit to find where the open or break in the circuit is. You need to set your meter for the appropriate range and then find an anchor point on the line 2 or neutral side of the circuit. For the example we were using, the anchor point would likely be on the far right of the compressor branch circuit just after the CS contact. We will call this the "anchored lead." You will move the other meter probe through the circuit from left to right. We will call this the "mobile lead." If you moved the mobile lead to a point between the CS contact on the left of the branch, between the contact and OL, and found near 120 VAC you know the CS contact is closed. Next you should move the mobile lead to the right side of the OL contact; the presence of 120 VAC would indicate that it is closed as well. If at either of these points you measured a voltage substantially lower than 120 or 0 VAC, that is an indication of a break in the circuit. As an additional note, if you found that the CS contacts were open, the solution is not to immediately condemn the compressor starter. It might be that the coil is not energized in the low voltage circuit; you would then need to hopscotch the circuit that the coil is in to see if 24 VAC is present at the CS coil. If there is 24 VAC present at the coil and the contacts are still open, the fault is with the relay and needs to be replaced.

PROCEDURE

Step 1. Using ED-01 we are going to go through a hop scotching process. Let's say we had a problem with the IFM. The point where we anchor one meter lead is at the black junction point on the right side of the IFM branch circuit. We take our mobile lead and place it on the left side of the IFM relay contact at the black junction point. What voltage *should* you measure?

Step 2. If you determine the correct voltage is available between the two points in Step 1, what *should be* the voltage when you move the mobile lead to the right side of the IFR contacts (terminal 4)?

Step 3. Using a trainer with a "bugged" control circuit that your instructor assigns you, use the same hopscotch approach to determine where the break in the circuit is and then list the fault you find in the space below.

QUESTIONS

1. Are there any special considerations when changing this component out?

2. With this control circuit failure, what would the symptoms be for the system? In other words, what would your service ticket say was the problem with the system?

3. After hop scotching a system you find that a pressure safety switch is open. What is the next step you should take?

4. If you measured the voltage on either side of a switch in a 24 VAC control circuit (with power applied) what will you find if the switch is closed (this is a shortcut version of hop scotching)?

Unit 10 Capacitors

Unit Summary

It is important to know the basic operation of a capacitor so that a good motor is not replaced due a lack of knowledge or experience. A capacitor can be used to start or run a motor, or both start and keep the motor running. A start capacitor creates a 90-degree current and voltage phase shift to create the maximum starting torque or starting turning power. A run capacitor allows the motor to run at a lower amperage condition plus offers a small amount of starting torque assistance. Single-phase compressors require a run capacitor connected in parallel to the run winding and in series with the start winding. This puts the two windings out of phase from one another and allows the compressor motor to start and run efficiently.

Capacitors are rated in microfarads and voltage. The voltage rating for a start capacitor should be equal to or greater than the starting voltage. The voltage rating for a run capacitor is about 100 volts higher than the supplied voltage. Standard voltage ratings on capacitors are 370 and 440 V. A higher voltage rating can be used, but using lower voltage rated capacitors will cause the capacitor to overheat and burn out.

Capacitors can be wired in series or parallel to, respectively, decrease or increase their microfarad rating. Calculating capacitance in series or parallel is the opposite of calculating resistance in series and parallel circuits. Knowing this information will allow you to get a piece of equipment running without having to run to the HVACR supply house.

When troubleshooting a capacitor, use a capacitor checker unless the capacitor is obviously damaged. The microfarad rating must be within 10% of the labeled rating to be considered good. Very few new capacitors are defective right out of the box. Test new capacitors for the correct microfarad rating prior to installing. Refer to Table 10-1 of your text for valuable information on the limits and applications for run and start capacitors.

Key Terms (Definitions can be found in the Glossary in your text.)

Bleed resistor	Microfarad
Capacitance rating	Run capacitor
Capacitor	Start capacitor
Dual capacitor	Starting torque
Farad	

LAB 10.1 Capacitor Selections and Testing

LABORATORY OBJECTIVE

In this lab you will learn how to determine the condition of a capacitor and the replacement rules

if you do not have an exact replacement.

ELECTRICITY FOR HVACR, 1e TEXT REFERENCE

Unit 10: Capacitors

REQUIRED MATERIALS PROVIDED BY THE STUDENT

- DMM or capacitor tester

REQUIRED MATERIALS PROVIDED BY THE SCHOOL

- 2 watt 15 ohm resistor for discharging capacitors.

- 8 non-dual capacitors, can be run or start or combination of the two.

- 8 dual capacitors

SAFETY REQUIREMENTS

Since capacitors can hold a charge, care should be taken to make sure all capacitors are

discharged before touching terminals (see Figure 10-1-1). Always use common sense and follow

all lab safety rules. As with any laboratory activity safety glasses should be worn at all times.

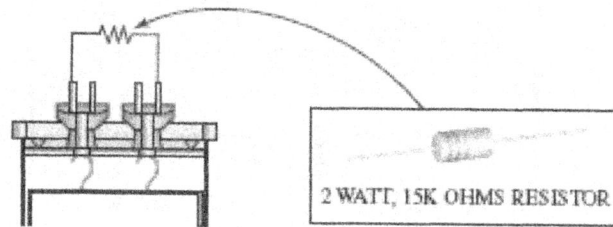

Figure 10-1-1. A bleed resistor across capacitor terminals. Always discharge capacitors before checking and discharge every time you check them.

INTRODUCTION

Many motors rely on capacitors for increased starting torque and improved running efficiency. Capacitors can be identified as running capacitors, which are in a metal case; a start capacitor, which is in a black plastic case; and a dual capacitor, which is two run capacitors in one metal case. The size of a capacitor is determined by the voltage or VAC and the microfarad rating, which can be abbreviated as MFD or μF. It is important to know how to identify capacitors as well as check the condition of them. Many times a visual inspection will let you know that the capacitor is bad. If the top or terminal end of a run capacitor is popped up like a dome, it needs replacing, see Figure 10-1-2. Since there are so many different microfarad ratings of capacitors, it is impossible for a service tech to have one of every size on his truck so the replacement rules are very important to know. The technician may not have the exact replacement on his or her truck, but may have one or two that can be used to replace a bad capacitor. This lab will introduce you to the procedure of checking capacitors and how to determine what size replacement capacitor you can use.

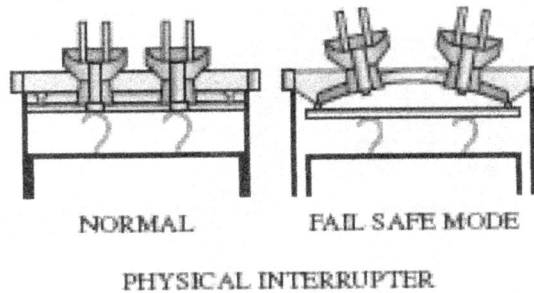

NORMAL   FAIL SAFE MODE

PHYSICAL INTERRUPTER

Figure 10-1-2. A capacitor with the dome popped up. This is a visual indication that this capacitor is bad and should be replaced.

Capacitor Replacement Rules

1. Voltage of replacement capacitor should be equal to or greater than the one it is replacing.

2. Microfarad of capacitor should be + or – 10% of the capacitor it is replacing.

3. Microfarad of start capacitor should be in the range listed on capacitor.

4. When capacitors are wired in parallel the microfarads of the two are added.

5. When capacitors are wired in series use formula CT =$\dfrac{C1 \times C2}{C1 + C2}$

See Figure 10-1-3 for an example of the necessary information found on a capacitor label.

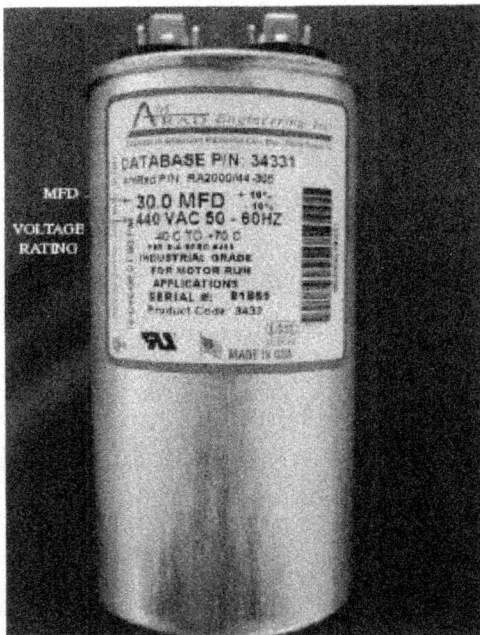

Figure 10-1-3. A run capacitor. Notice the voltage rating of 440 VAC and microfarad of 30. Microfarads can be abbreviated as MFD, µF or cap.

PROCEDURE

Step 1. Your instructor will assign you eight non-dual capacitors to evaluate. It is very important that you discharge each with a 2 watt, 15K ohm resistor before checking and between checks; see Figure 10-1-1. If any capacitors have a resistor attached it must be removed for accurate readings. Record the type of capacitor, the voltage, and the rated μF of each. If you have a capacitor tester, check and record the capacitance and conditioning of each. If you only have a DMM, you cannot determine the actual μF so just record the condition of each.

Type Start or run	Voltage	Rated μF	Actual μF	Condition Open, short, good

Step 2. Your instructor will assign you eight dual capacitors to evaluate. It is very important that you discharge each with a 2 watt, 15K ohm resistor; see Figure 10-1-1. This lab is to get you familiar with the capacitor replacement laws. You will record the voltage and the voltage range of the replacement, meaning if voltage is 100 and the law states "equal or greater," any voltage

over 100 VAC is fine. Record the fan MFD and herm MFD; see Figures 10-1-4 and 10-1-5. Then

record the range of MFD for the replacement capacitor using + or – 10% for run capacitors. For

voltage range and capacitance range see replacement rules above.

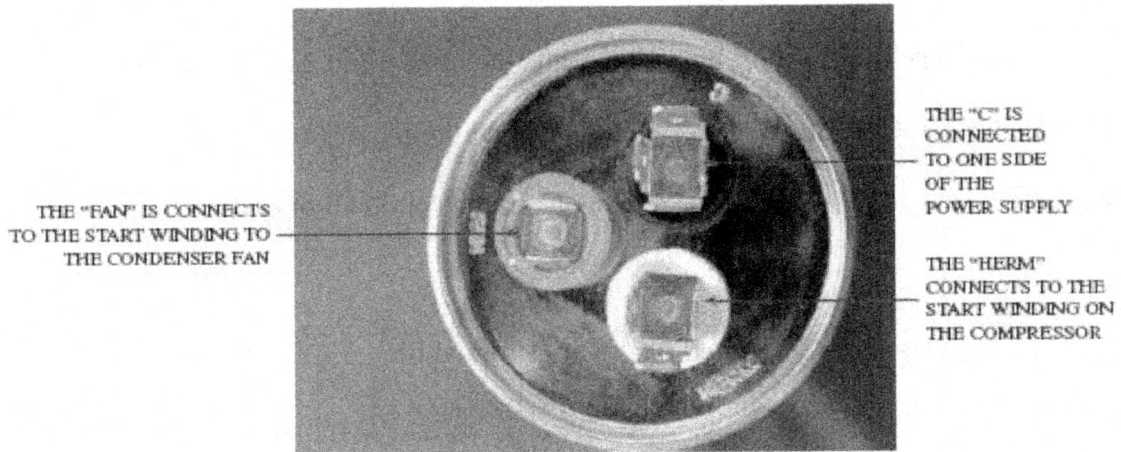

THE "FAN" IS CONNECTS
TO THE START WINDING TO
THE CONDENSER FAN

THE "C" IS
CONNECTED
TO ONE SIDE
OF THE
POWER SUPPLY

THE "HERM"
CONNECTS TO THE
START WINDING ON
THE COMPRESSOR

Figure 10-1-4. A dual capacitor. There are three connections on a dual capacitor, but only two capacitors. C o(r common) and "FAN" are considered one capacitor, and C (or common) and "HERM" is the other.

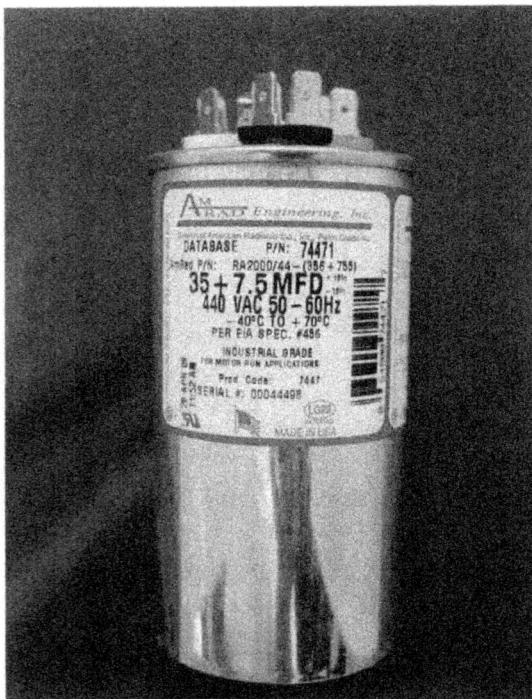

Figure 10-1-5. The label of a dual capacitor. The 35 is the microfarad capacitance of the herm side and 7.5 is the microfarad capacitance of the fan side of this capacitor. The voltage (or VAC) 440 is the same for both the fan side and herm side of the dual capacitor.

Voltage	Voltage range	Rated µF		Replacement capacitance range	
		Fan	Herm	Fan	Herm

QUESTIONS

1. A 25 µF and a 35 µF capacitor are wired in parallel. What is the total capacitance of the two?

2. A 25 µF and a 35 µF capacitor are wired in series. What is the total capacitance of the two?

3. You need to replace a capacitor that is 370 VAC 50 µF. Can you use one that is 440 VAC?

4. What is the capacitance range for the above question?

5. If a run capacitor has a marked terminal, the marked terminal is connected to what?

6. A start capacitor has a microfarad range on the nameplate, if its actual μF is 10% below this range is the capacitor good?

7. What type of motor would use a start and run capacitor?

8. Why should you not discharge a capacitor with a screwdriver?

Unit 11 Thermostats

Unit Summary

A thermostat is merely a switch that automatically controls a heating, cooling, or refrigeration system. The modern digital thermostat is more accurate, easier to read, and controls the setpoint temperature more closely than do the old-style mechanical thermostats. Some models are programmable. To save energy, the contractor or homeowner can set times of the day when the equipment can be turned up or down. Some digital thermostats actually notify the user that there is a system problem. The thermostat can notify a customer or their contractor by phone or via the Internet. The thermostat can transmit a fault indicator so that the technician will have an easier time finding the problem.

The common thermostat terminals are **R**, **Y**, **W**, and **G**. The **C**, or common terminal, is found on some thermostats. The **R** terminal is a red wire and is the power from one side of the transformer. The **Y** terminal is yellow and sends power to the cooling contactor coil. The **W** terminal is white and sends power to the heating control. The **G** terminal is green and controls the indoor blower operation. The **C** terminal is the other side of the 24-V transformer and is required to complete the circuit for a digital thermostat or to operate heating and cooling controls.

Finally, troubleshooting a thermostat is not difficult. First, do a system survey using the ACT method or some other logical method of troubleshooting. If the solution to the problem leads back to the thermostat, remove the thermostat and, from the subbase, jumper from **R** to **Y**, **R** to **W**, and **R** to **G**. Think of the jumper sequence as a manual thermostat. The fan, cooling, and heating circuits should work when jumping. The cooling, heating, and blower should all work. If one of the circuits checked does not work, the problem is that specific thermostat wire or that specific control.

Key Terms (Definitions can be found in the Glossary in your text.)

Cooling anticipator	Setback temperature
Deadband	Setpoint
Digital thermostat	Setpoint temperature
Heat anticipator	Snap action thermostat
Humidistat	Solid state thermostat
Line voltage thermostat	Subbase
Mechanical thermostat	Thermistor
Programmable thermostat	Thermostat

LAB 11.1 Mechanical Thermostat Function and Selection

LABORATORY OBJECTIVE

You will be able to properly select the type of mechanical thermostat needed for a system and correctly wire the thermostat.

***ELECTRICITY FOR HVACR, 1e* TEXT REFERENCE**

Unit 11: Thermostats

REQUIRED MATERIALS PROVIDED BY THE STUDENT

- 6-in-1 screw driver

- Wire strippers

- DMM or VOM suitable for HVACR field work

REQUIRED MATERIALS PROVIDED BY THE SCHOOL

- Heat / Cool mechanical thermostat mounted on a board

- Heat Pump mechanical thermostat mounted on a board

- HVAC system of instructor choice to wire thermostat

- Thermostat wire and small wire nuts

SAFETY REQUIREMENTS

Follow all lab and shop safety rules, which includes wearing safety glasses at all times. Adhere to lockout/tagout procedures and never assume power is off, always check with a multimeter. It is unsafe to work around live electricity wearing a watch or rings, so remove them before starting this task. The switching component in most mechanical thermostats is mercury, which is very hazardous. Care should be taken when handling these thermostats so they do not break and leak mercury. If this happens, notify your instructor immediately.

INTRODUCTION

This lab is about a type of thermostat that uses a mechanical means to control the temperature of a space by opening and closing switches inside the thermostat. Mercury switches and magnetic switches (see Figure 11-1-1) are the two main types of temperature-controlled switches used in mechanical thermostats. The opening and closing of these switches redirects the low voltage to the controls in the system that need to operate. A typical mechanical thermostat will have a mode switch that can be set on heat, cool, off, or auto. The auto setting means the thermostat can automatically change over from heat to cool as the temperature changes. A fan switch set to "auto" means the indoor fan will run with a call for heat or cool. A fan switch set to "on" means the indoor fan will run continually. The thermostat will also have some means of setting the desired temperature and a thermometer on the front so the homeowner has reference for the current temperature. It is very important that mechanical thermostats be mounted level so the mercury switch will operate at the correct temperatures.

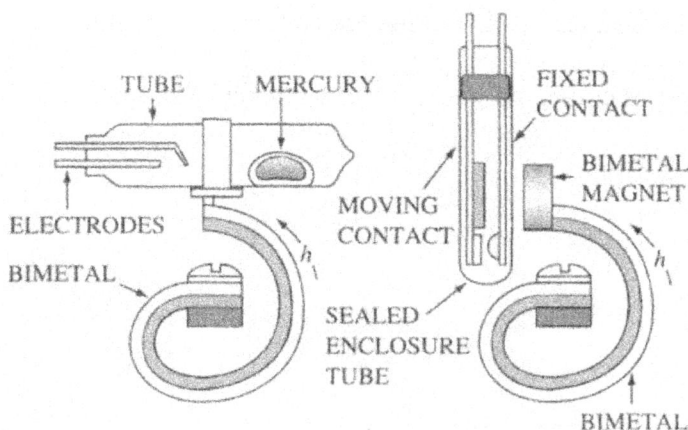

Figure 11-1-1. The thermostat element on the left is a mercury filled glass bulb. The snap action thermostat on the right closes when the magnet gets close to the moving contact.

PROCEDURE

Step 1. Mount a mechanical heat/cool thermostat subbase on a board. The subbase should be level and the board should be secure. Wire the correct subbase terminals with a short piece of thermostat wire. To determine the correct wire color, refer to Table 11-1, "Summary of Terminal Connections," in your text. Attach the thermostat to the subbase. Test resistance on the wires connected to terminals with your ohmmeter as you check the different functions of thermostat. In the table below you will be checking the resistance or continuity between the red wire or R terminal, which is power to the thermostat, and other terminal wires that have different functions. On the first line of the table you will check between the R and G terminal or Red wire and Green wire. Record only closed (0 resistance) or open (OL) in boxes corresponding to the procedure. Closed would mean that low voltage power will be sent from R to other devices in control circuits to be energized. The fan switch has two settings, auto or on. "On" means the indoor fan will run all the time; "Auto" means the indoor fan only runs with a call for Heat or Cool. Mode: Cool/Call for cool means the thermostat is in the cooling mode and the temperature is set lower than room temperature.

Mode: Heat/Call for heat means the thermostat is in the heating mode and the temperature is set higher than room temperature. The first line in the exercise is completed as an example, note that the type of heater will determine if the indoor fan comes on with Mode: Heat/Call for heat.

T-Stat terminals	Mode: off Fan: auto	Mode: off Fan: on	Mode: cool Call for cool Fan: auto	Mode: heat Call for heat Fan: auto
R - G (example)	open	closed	closed	open
R - G				
R - Y				
R - W				
R - C				

Step 2. Mount a mechanical heat-pump thermostat subbase on a board. The subbase should be level and the board should be secure. Wire the corresponding subbase terminals with a short piece of thermostat wire. To determine correct wire color, refer to Table 11-1, "Summary of Terminal Connections," in your text. Attach the thermostat to the subbase. Test resistance on the wires connected to terminals with your ohmmeter as you check the different functions of the thermostat. In the table below you will be checking the resistance or continuity between the red wire or R terminal, which is power to the thermostat, and other terminal wires that have different functions. For the first line of the table you will check between R and G terminal, or Red wire and Green wire. Record only closed (0 resistance) or open (OL) in boxes below corresponding to the procedure. Closed would mean that low voltage power will be sent from R to other devices in the control circuits to be energized. The fan switch has two settings: auto or on. "On" means the indoor fan will run all the time; "Auto" means the indoor fan only runs with call for Heat or Cool. Mode: Cool/Call for cool means the thermostat is in the cooling mode and the temperature

is set lower than room temperature. Mode: Heat/Call for heat means the thermostat is in the heating mode and the temperature is set higher than room temperature.

T-Stat terminals	Mode: off Fan: auto	Mode: off Fan: on	Mode: cool Call for cool Fan: auto	Mode: heat Call for heat Fan: auto
R-G				
R-Y				
R-W				
R-O				
R-B				
R-C				

Step 3. Your instructor will assign you a system. You will first decide which thermostat needs to be installed by determining what type of system you've been assigned, for instance gas heat, air conditioning system, heat pump, etc. Once the correct thermostat has been determined, mount the subbase 60 inches (five feet) above the floor and level it. Wire the subbase and have the instructor check the connection. Care should be taken to ensure no bare wires are touching each other and that wires are wrapped around the subbase screws in the correct direction, meaning the direction that tightens the screw. Once the subbase is properly wired, mount the thermostat to the subbase and check system functions.

Type of system to install thermostat on _____

Check thermostat functions in the order listed below with your instructor's supervision.

_____ Mode off, Fan on – indoor blower should come on.

_____ Mode off, Fan auto – indoor blower should turn off after a short delay.

_____ Cool mode on, call for cool – the air conditioner and indoor blower should come on.

_____ Cool mode on, no call for cooling – the air conditioner and indoor blower should turn off.

_____ Heat mode on, call for heat – the heat and indoor blower should come on after a delay.

_____ If you are working with a gas heating system, set the heat anticipator to the proper setting.

   (Refer to Section 11.8 in your text for the procedure.)

_____ Heat mode on, no call for heat – the heat and indoor blower should turn off.

_____ If you are working with a heat pump thermostat: thermostat to emergency heat, call for

heat – the electric heat and indoor blower should come on.

QUESTIONS

Match the following thermostat functions with the thermostat terminals.

_____ 1. reversing valve energized in cooling mode   A. B

_____ 2. transformer common   B. R

_____ 3. 1st stage compressor contactor   C. Y

_____ 4. indoor fan   D. W

_____ 5. 2nd stage heat   E. C

_____ 6. reversing valve energized in heating mode   F. Y2

_____ 7. 2nd stage compressor contactor   G. G

_____ 8. transformer power   H. O

_____ 9. 1st stage heat   I. W2

      J. X

LAB 11.2 Electronic and Programmable Thermostat Selection and Function

LABORATORY OBJECTIVE

You will learn how to install and operate electronic thermostats and how to install, operate, and program programmable thermostats.

***ELECTRICITY FOR HVACR, 1e* TEXT REFERENCE**

Unit 11: Thermostats

REQUIRED MATERIALS PROVIDED BY THE STUDENT

- 6-in-1 screw driver
- Wire strippers
- DMM or VOM suitable for HVACR field work

REQUIRED MATERIALS PROVIDED BY THE SCHOOL

- A residential heating and cooling system with installation instruction
- A battery back-up electronic thermostat with installation instruction
- A programmable thermostat with installation instruction

SAFETY REQUIREMENTS

Follow all lab and shop safety rules, which includes wearing safety glasses at all times. Adhere to lockout/tagout procedures and never assume power is off, always check with a multimeter. It is unsafe to work around live electricity while wearing a watch or rings, so remove them before starting this task. Care should be taken not to wire thermostats with low voltage power on; this can be a shock hazard and could short out transformer if wires touch. Always make sure a low voltage fuse is in place to protect the transformer.

INTRODUCTION

This lab is designed to familiarize you with the selection and installation of electronic and programmable thermostats. Mechanical thermostats were the main type of temperature control for a long time. The mercury in these thermostats is hazardous to the environment and special precautions must be taken to dispose of them. Electronic thermostats are much more accurate at controlling temperature and can be disposed of without environmental concerns. Electronic and programmable thermostats have functions not found in mechanical thermostats, like outdoor temperature, dirty air filter, and current time and date display. It is now possible to have a Web-enabled thermostat that you can set from your cell phone. Many states have adopted the required use of programmable thermostats for new home construction into their state energy code as a way of lowering the operating cost to the homeowner. Electronic and programmable thermostats are not the way of the future, they are the present, and technology is changing in them every day.

PROCEDURE

Step 1. Your instructor will assign you a residential heating and cooling system to wire. It is possible that you will need to wire the line voltage and low voltage to the equipment. If you cannot remember how, then this is a good refresher. The instructor will give you an electronic thermostat, with installation instruction, to install with equipment. The thermostat needs to be one with a battery pack inside. The first thing to decide is the type of equipment, whether it is gas heat, heat pump, etc. This will determine how the thermostat is wired.

What type of equipment have you been assigned?

Wire the equipment and install the thermostat per installation instructions.

Does the digital thermostat have a battery pack?

If "yes," you will continue work with the battery pack in Step 2.

Have your instructor check your wiring.

Check thermostat operation in this order:

- Mode off, Fan on – indoor fan should run.

- Mode off, Fan auto – indoor fan should turn off.

- Mode cool, Call for cool – indoor fan and air conditioning should come on.

Note: Electronic thermostats typically have a built-in time delay, which is usually 5–10 minutes.

- Mode cool, No call for cool – air conditioner should turn off, indoor fan off after delay.

- Mode Heat, Call for heat – indoor fan and heater on.

- Mode Heat, No call for heat – heater should turn off, indoor fan off after delay.

Will this thermostat alert homeowner when air filter is dirty?

How does it detect when air filter is dirty?

Step 2. Some electronic thermostats have a battery pack so they can be installed without a low voltage common wire. If an air conditioning tech does not have enough wires to wire the thermostat and no easy way of pulling another wire, a battery back-up thermostat is necessary. When there are not enough wires to properly wire the digital thermostat, a battery pack thermostat can be used instead of connecting the C terminal in the thermostat. The batteries are used to light the digital display instead of the 24 volts, which typically comes from R and C. If your thermostat has a place for batteries, unwire the C terminal and install batteries and complete the following checklist:

Does the digital display on thermostat work?

Does equipment work properly?

Now take batteries out of thermostat.

Does the digital display on thermostat work?

Does equipment work properly?

Step 3. Your instructor will assign to you a residential heating and cooling system to install a programmable thermostat on. The programmable thermostat can replace the electronic thermostat in Step 2. When the programmable thermostat is wired, have your instructor check your wiring. The first step in programming is to set the correct date and time, and then proceed to program the provided schedule into the thermostat.

	Time	Cool temp	Heat temp	Weekend settings	
				Cool temp	Heat temp
Wake	7:00 am	78	70	80	65
Day	9:00 am	85	62	82	68
Evening	5:00 pm	76	70	76	70
Sleep	11:00 pm	80	68	82	68

Have your instructor check your program when you have completed it.

Show your instructor how to bypass the program with the hold setting.

QUESTIONS

1. How does a digital thermostat detect temperature?

2. What is the purpose of the override in a programmable thermostat?

3. Why do some digital thermostats have a battery pack in them?

4. Is it necessary to level a digital thermostat?

5. The temperature that is programmed into a thermostat to maintain a temperature when the homeowner is not home is called what?

6. The setting of a programmable thermostat for when the homeowner is at work is called?

7. A programmable thermostat in which every day of the week can be programmed differently is called a _____ day thermostat.

8. How long is the normal time delay of most digital thermostats?

Unit 12 Pressure Switches

Unit Summary

This unit discussed some of the common pressure switches used in air conditioning and refrigeration equipment. Other types of pressure switches are also used in various heating, air conditioning, and refrigeration applications.

The low-pressure switch is used to temporarily break power to a circuit when the suction drops below a preset level. Low refrigerant pressure will cause a compressor motor to overheat, since the volume of the cool gases is reduced. Low refrigerant can also cause the evaporator coil to freeze up, thus reducing heat transfer and airflow. The ice layer can become several inches thick and will need to be thawed prior to operating the system again.

The high-pressure switch is used to temporarily or permanently stop the operation of a compressor. The temporary open circuit is called an automatic reset, which means that the high-pressure switch will close when the pressure drops to a preset level. The high-pressure switch with a manual reset will require the tech to push a button on the switch that closes the contacts and restarts the compressor. It is important for the compressor to be protected from high-pressure and overheating conditions. These conditions will severely damage the compressor if it is left to operate in this high-pressure condition.

The fan cycling switch is used to turn off the condenser fan when the head pressure drops too low to effectively push liquid refrigerant through the metering device. Most metering devices require a 100-psi pressure differential between the high side and low side in order to achieve the required refrigerant flow rate to cool properly.

The oil pressure safety switch keeps the compressor from being damaged due to a lack of lubrication. The oil safety switch operates on a pressure differential. This is called *net oil pressure*. The oil safety switch operates on the pressure difference between the suction pressure and oil pump pressure. Before the compressor starts, the suction pressure and oil pump pressure are the same. As the compressor starts, the suction pressure drops and the oil pump pressure increases. The oil safety switch must develop at least a 9-psi differential in 120 seconds of running time. If the differential is not developed in the allotted time, the oil safety switch will open the control circuit and stop the compressor.

This unit also discussed the installation and troubleshooting process required of these pressure switches. It is important to check the "failure" mode operation of the switch when it is replaced. The replacement installation will require that the switch be checked to ensure it has been wired correctly. It is a "best practice" to check any replacement part under the conditions in which it will be operating. If it is a safety device, it should be tested under the conditions it was designed to protect against; for example, you could create a low-pressure condition to see if the low pressure stops the compressor when the suction pressure drops below its setpoint.

Key Terms (Definitions can be found in the Glossary in your text.)

Cut-in   Fan cycling switch

Cut-out   Ground fault current interrupter

Electronic expansion valve   High-pressure switch

LAB 12.1 Pressure Switches and Other Safety Switches Used in HVACR

LABORATORY OBJECTIVE

The purpose of this lab is to familiarize you with common pressure and safety switches that are used in HVACR systems. You will learn the different types of switches that are available, how they operate, and how to select appropriate switches for a particular application.

***ELECTRICITY FOR HVACR, 1e* TEXT REFERENCE**

Unit 12: Pressure Switches

REQUIRED MATERIALS PROVIDED BY THE STUDENT

- DMM or VOM suitable for HVACR field work

REQUIRED MATERIALS PROVIDED BY THE SCHOOL

- Four different types of refrigerant pressure switches

- Four different types of temperature limit switches

- One manual reset pressure switch

- Pressure testing gas (air or nitrogen)

- Pressure switch reference catalog

SAFETY REQUIREMENTS

You will be using air or nitrogen pressure to test various switches so proper safety equipment must be worn at all times. Follow all safe hand-tool use practices and use caution when using pressure or hot air guns to test devices. Make sure the device you're testing is rated for handling the temperature or pressure extreme you are exposing it to before beginning.

INTRODUCTION

HVACR systems are by their very nature related to pressure and temperature. When dealing with heating systems there are always controls and switches installed by manufacturers that protect against overheating and fires. In many cases there are multiple protection devices. Many gas-fired appliances contain roll out switches to shut the burners down if a flame escapes the combustion zone, and some also have a temperature-limiting switch that also shuts the burner down if the heat exchanger overheats. Cooling and refrigeration systems can be damaged and cause injury to bystanders if the high-pressure side of the system becomes dangerously high so these switches are often used to prevent over-pressure conditions. Low-pressure switches protect against low side pressure becoming too low and damaging equipment.

There are no HVACR systems that do not contain one or more safety switches in their design and it is important to be able to identify these important devices and select the appropriate one to replace a failed switch or install a new system.

Temperature safety switches can be either manual or automatic reset devices. Manual reset devices are used to keep the system shut down in the event of a more serious problem until a technician examines it before the equipment is allowed to restart. Flame roll out switches are often manual reset devices, although they can also be one-time devices that open once and must be replaced after opening, much like a fuse. Temperature safety switches sense temperatures that are too high or too low. The settings of the switch (cut-in, cut-out, and differential) are based on the type of system and what the manufacturer calls for.

Pressure safety switches are similar in that they can be manually reset or automatically reset. Pressure switches can respond to any pressure, refrigerant, air, or water.

All of these switch types come in many arrangements and styles. They can vary from SPST all the way to DPDT type switches. They can have adjustable settings or they can have a fixed cut-in, cut-out or differential. The mounting style can be more universal or they can be make or model specific. Field experience is the best training tool for determining which type of switch is best to use for each application, but you will learn some of the basics in this lab assignment.

PROCEDURE

Step 1. Complete the table using the four refrigerant pressure switches selected by the instructor.

Pressure Switch Number	Pressure the Switch Cuts-out at	Pressure the Switch Cuts-in at	Range (N/A if a fixed switch)	Type of Pressure Switch (High or Low Side)	Is this a Fixed or Adjustable switch?
1					
2					
3					
4					

Step 2. Complete the table using the four temperature safety switches selected by your instructor.

Temperature Switch Number	Temperature the Switch Cuts-out at	Temperature the Switch Cuts-in at	Range (N/A if a fixed switch)	Application of Switch (Roll out, freeze protection etc.)	Is this a Fixed or Adjustable switch?
1					
2					
3					
4					

Step 3. With the manual reset switch your instructor selected, determine at what pressure this switch opens its contacts (cut-out). Record this value below.

_____ psig

Step 4. Now you are going to bench test this switch to see if it cuts-out or cuts-in at the pressure it is supposed to. After getting approval from your instructor, connect the pressure switch capillary or flare connection to a nitrogen or compressed air line and slowly increase the pressure until the contacts open (or close if a close on pressure rise type). You will want to have a manifold gauge set connected to the switch so you can closely monitor pressure. If the pressure exceeds the specified value and does not open or close you should stop increasing the pressure; the switch has failed. Verify the switch has opened or closed by testing for continuity with a DMM.

Step 5. If the switch trips successfully, try resetting the switch with the manual reset. Use a DMM to make sure the switch has reclosed.

QUESTIONS

1. A manual reset switch will always be a safety type control. Why?

2. The manual reset pressure switch you looked at will trip when the pressure limit is reached. What problems in a system might cause this switch to trip (other than the switch failing)?

3. The switches you examined for this lab were all safety devices. List a few examples of switches that monitor temperature or pressure that are *operating* controls and not used for safety reasons.

4. Imagine you are installing an air conditioning system and that you want to install a high-pressure switch on for safety. This is a traditional split system and is using R-22. Using the catalog your instructor provides, look up an appropriate high-pressure safety and list the part number for it below.

5. You have a job working on an older electric heating system for which parts from the original manufacturer are no longer available. You find that the high temperature limit switch has failed and needs replaced. The label on the old limit shows the cut-out is 200°F. Using the parts manuals your instructor provides, locate a replacement high limit switch and record the part number below.

6. All the safety switches we looked at monitor temperature or refrigerant pressure. What other things might a safety switch monitor to detect a safety issue?

Unit 13 Miscellaneous Electrical Components

Unit Summary

This unit discussed common electrical components used in HVACR equipment. Understanding these common components will help in the troubleshooting process. There will always be electrical components and refrigeration systems that you will not understand. When this occurs, review the component's features and its position in the electrical system. This may help you determine the component's function. If the undetermined component is in the working part of the circuit, ignore its operation. If the undetermined component is in the nonworking part of the circuit, you may need to call for help. Calling for help is not a sign of weakness, since no one technician can possibly be familiar with all electrical components, especially with all of the new components in use today.

The first component covered was the crankcase heater. The heater is important for keeping the refrigerant warm. The heat keeps the oil warm and prevents the refrigerant from condensing in the compressor crankcase.

A solenoid valve is an electromechanical valve that opens or closes by energizing a magnetic coil. The solenoid valve is commonly found in a liquid line, including water and other fluid lines. Electric unloaders are used on semi-hermetic compressors to reduce the capacity of a compressor when lower capacity operation is required. This allows the compressor to continue running without cycling off.

A solid-state timer will delay the operation of a compressor for about 5 minutes. This time delay allows the pressures to equalize and the power quality to stabilize.

The lockout relay is a type of device that is usually found in a control circuit. As the name implies, the lockout relay prevents or locks the compressor circuit from operating when a safety device opens. The lockout circuit is reset when the power is turned off and back on.

A hot gas sensor is designed to keep compressors from overheating and burning out. The discharge line will become excessively hot when there is a low refrigerant charge or inadequate compressor lubrication.

The line voltage monitor provides many protective features. The monitor protects against short cycling, voltage imbalances, phase reversals, and low-or high-voltage conditions.

The unit discussed explosion-proof designs. It is important to be able identify explosion-proof designs and be able to access the enclosed wiring and connections. Equally important is making sure you seal the enclosure back to its original condition so that it does not create a danger to the surrounding environment.

The final section talked about internal and external overload devices. These safety devices are used to protect the motor from high amperage conditions that could burn out a motor winding or cause the motor shell to get hot enough to ignite a fire.

Key Terms (Definitions can be found in the Glossary in your text.)

Electric unloader	Holding relay
Explosion-proof systems	Impedance relay

Line voltage monitors

Lockout relay

Mechanical unloader

Reset relay

Solenoid valve

Solid state timer

Under load

Unloaders

Unit 14 How Motors Work

Unit Summary

The focus of this unit was to help you understand how motors work. We discussed how magnetism is used to get a motor moving in the right direction. Single- and three-phase motors are used in our industry. The nameplate will tell you the voltage type and many other pieces of information to determine motor operation and motor capabilities. It is important that you understand all of the information provided on the nameplate, but especially the voltage source, maximum amperage capabilities, RPM, shaft size and rotation, and frame number. Learning the many features of a motor is beneficial when it comes to troubleshooting and replacing motors. Most techs do not realize what goes into a motor's design until it fails and they are not able to obtain an exact replacement. The air conditioning supply houses or motor suppliers can help you find the correct replacement. In some cases a motor is so specialized that only the original equipment manufacturer's (OEM) motor will do. Defective motors can be rewound and bearings replaced, but that usually takes a few days.

Twenty-five percent of motors returned under warranty are not defective. What are technicians thinking when they install a new motor and it does not work? Do they think they have a new defective motor? It could happen, but the chances are very slim. Try to check out the motor before it is totally installed. At least check the resistance of the windings. Damage can occur during shipping and it is best to know this before the motor is installed. Actually, checking the resistance at the supply house is the best practice, but is rarely done.

Replacing starting components will ensure long motor life. A weak capacitor or erratic starting relay may not be apparent when a new motor is replaced. The short life of the new motor will have you double checking the starting components. Yes, there will be a motor replacement warranty but it may not cover travel and labor costs. Some warranties require starting component replacements also.

Key Terms (Definitions can be found in the Glossary in your text.)

Babbitt lined bearings	Multispeed motors
Ball bearings	National Electrical Manufacturers Association (NEMA)
Frame size	
Frequency	Power factor (PF)
Horsepower	Permanent split capacitor (PSC) motor
Insulation class	
	Rated voltage
Magnetism	
	Revolutions per minute (RPM)
Motor nameplate	

Rotor

Run capacitor

Self-aligning sleeve bearing

Service factor

Shaded pole motor

Single-phase motor

Slip

Start capacitor

Starting torque

Stator

Synchronous speed

Unit bearings

Unknown soldier

LAB 14.1 Introduction to Motor Types

LABORATORY OBJECTIVE

The purpose of this lab is to familiarize you with the many types of motors used in HVACR systems. Each motor type is used for a specific purpose and you will learn how to identify each based on nameplate ratings, the equipment they are installed on, and the accessories that are found with the motor.

ELECTRICITY FOR HVACR, 1e TEXT REFERENCE

Unit 14: How Motors Work

Unit 15: Motor Types

REQUIRED MATERIALS PROVIDED BY THE STUDENT

- DMM or VOM suitable for HVACR field work

REQUIRED MATERIALS PROVIDED BY THE SCHOOL

- One of each type of motor installed in a system and numbered (Shaded Pole, Split Phase, CSIR, PSC, CSR, and Three Phase)

- Motor replacement catalog or website

SAFETY REQUIREMENTS

This lab may involve applying high voltages to motors and capacitors. You should discharge all capacitors before handling them with an appropriate capacitor-shorting resistor. Practice all lab safety precautions you have in the past when dealing with high voltages. This lab also demands special attention since we are working with motors that rotate very quickly. It is very easy to get a finger or other part of your body wrapped in a motor or fan blade, be very careful when working around operating motors. Don't leave any tools or parts near an operating motor, they can be thrown out at a very high speed.

INTRODUCTION

Motors are found in virtually every HVACR system with few exceptions. With the wide variety and sizes of systems available there must also be a wide variety of motors available to match each application.

Motors have many ratings and specifications as described in Unit 15 of your textbook. Your job for this lab is to identify all motor types based on these ratings, the equipment it is found in, and the accessories that are found on the motor. You will see capacitors, starting relays, and overload devices found on many of these motors. These items give you a clue as to which type of motor you are looking at. Use the following descriptions to help you identify motor types for this lab.

Shaded pole motors: These have the lowest starting torque and running efficiency of any motor type. They are used for small appliances and fans where efficiency is not a concern. They are usually inexpensive to purchase. They do not contain any special components and will be a fractional horsepower motor. Examples of shaded poles motors are found in Figure 14-1-1.

Figure 14-1-1. This image shows two examples of shaded pole motors. The left one would have an enclosure and be more difficult t distinguish from other types of motors. The motor on the right is a skeleton or C-frame shaded pole motor.

Split phase (RSIR) motors: RSIR motors will always have some type of starting relay. They use either an internal centrifugal switch or current type starting relay (see Figure 14-1-2). They have a higher starting torque than a shaded pole motor, but it is lower than most other types. The running efficiency is fair. RSIR motors contain two windings; the start winding and run winding.

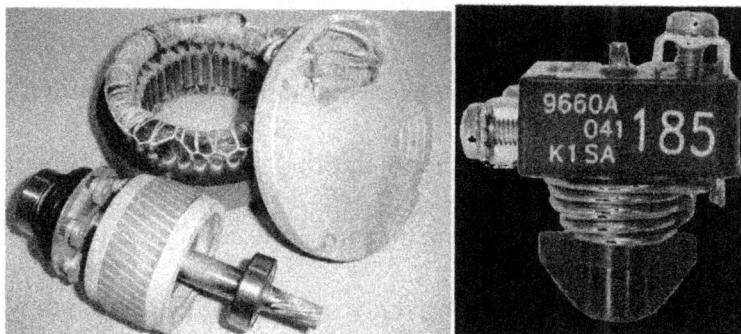

Figure 14-1-2. This image shows the types of starting relays usually found on RSIR type motors. On the left is a centrifugal switch mounted on the end of the rotor and on the right is a current relay.

CSIR motors: These motors are characterized by having a starting relay and a starting capacitor to provide extra starting torque. They are used for larger motors like those found in air conditioning and refrigeration compressors and in some pumps and fan motors. The start capacitor is always external and can be easily recognized by a black plastic case (usually) and a microfarad (μf) rating of about 60μf or higher. CSIR motors also contain a start winding and run winding. CSIR motors also require the use of a starting relay, like a current relay, centrifugal switch, or a potential relay (see Figure 14-1-3).

Figure 14-1-3. This is a picture of a potential relay that might be used on a CSIR or CSR motor.

PSC motors: PSC motors are commonly found in heating and cooling systems due to their better efficiency and lower amperage draw. They are a lower starting torque motor, though, since they contain no start capacitor, only a running capacitor. Running capacitors are usually made with a metal enclosure and usually have a microfarad rating of 60µf or less. The PSC motor also uses both a start winding and run winding.

CSR motors: These are heavy-duty, single-phase motors that also use both a start and run winding. They also use both types of capacitors: starting and running. This gives this type of motor a high starting torque and low running current draw compared to other single phase motors without these. There is also a starting relay used on CSR motors; they can be current, potential, or centrifugal switch types.

Three phase motors: Three phase motors are used in commercial and industrial equipment applications. They are unique since they do not require the use of capacitors or starting relays. They can be identified by the absence of these components. Three phase motors will, however, always contain a contactor or motor starter. These motors have very high starting torque and excellent running efficiency with a relatively low current draw.

PROCEDURE

Step 1. Using the six numbered motors that are installed in various HVACR systems your instructor selects for you, fill in the table with information about each. After you identify the components found on the motor you should know its type.

Motor #	1	2	3	4	5	6
Single Phase or Three?						
Start Capacitor? (Y/N)						
Run Capacitor? (Y/N)						
Starting Relay? (Y/N)						
Type (Split phase, shaded pole, etc.)						

Step 2. Identifying specific characteristics about a motor will help you identify problems as well as select an appropriate replacement motor. Using the same six motors, identify the following ratings and characteristics. Use Units 14 and 15 of your text for an explanation of each characteristic. This is not an exhaustive list but will get you started. You will need as much information as possible to find a replacement motor.

Motor #	1	2	3	4	5	6
Voltage Rating						
RLA						
LRA						
Rotation Direction						
NEMA Frame #						
RPM						
Shaft Diameter						
Horsepower						

Step 3. After getting approval from your instructor, start each of these six motors and observe their operation. You may want to make some mental notes about how quickly each one starts and shuts off. Do they start slowly and coast to a stop or do they start and stop abruptly like a car engine would? Pay attention to how each motor sounds when starting, stopping, and running. These can be hints that are useful for finding problems in the field.

QUESTIONS

1. Identify a possible application for each of the six motors you used for this lab. Indicate what type of load it might drive, such as a fan, pump, or compressor, and what type of environment you would expect to find it in (residential, commercial, or industrial).

2. Other than the characteristics you listed in the second step of your lab assignment, what others might be important? Use your textbook for some ideas.

3. Using one of the six motors you examined, try to locate a replacement from the catalog or website provided by your instructor. List the part number or stock number below.

4. If you replaced a correctly sized CSIR motor with a PSC type motor (which would be incorrect) what symptoms do you think you would see? What measurements might you take to indicate this is not the correct motor?

LAB 14.2 Motor Control Systems and Diagrams

LABORATORY OBJECTIVE

As we discovered in Lab 14.1, there are several types of motors used in HVACR systems. In this lab we will be looking at the details about controlling and operating each of these motors. There are many control schemes that can be used, depending on the type of system the motor is found in. Some motors are very easy to control while others have very complex wiring. Motor manufacturers have claimed that 25% of motors returned to them have "no problem," which means the fault was likely in the control system so it is very important for an HVACR technician to be able to wire and diagnose the control system.

ELECTRICITY FOR HVACR, 1e TEXT REFERENCE

Unit 14: How Motors Work

Unit 15: Motor Types

REQUIRED MATERIALS PROVIDED BY THE STUDENT

- None

REQUIRED MATERIALS PROVIDED BY THE SCHOOL

- Shaded pole fan motor trainer or system

- Capacitor start motor trainer or system

- Three-phase motor trainer or system

SAFETY REQUIREMENTS

In this lab we will not be energizing the motors that are being examined, but you should still use caution around equipment that is turned off. Lockout/tagout procedures are very important. You should follow all safe hand-tool use practices.

INTRODUCTION

In addition to the starting components that you studied for Lab 14.1, there are other external components that motors require to operate. Most motors in HVACR systems are automatically controlled by some condition, which means that there must be automatically operated switches to control them. Motors can be switched on or off by any of the methods we have discussed in the past; pressure, temperature, and liquid level are all common examples. There are not many examples of motors that require personnel to operate manually.

For many HVACR systems there is more than one motor that will need to be operating at different times, e.g., a traditional split air conditioning system. There is a minimum of three different motors that operate at different times in this system: the compressor motor, the condenser fan motor, and the evaporator blower motor. The compressor and condenser fan often run at the same time and they are operated by one control system. The evaporator blower motor needs to operate independently of the other motors and usually has a completely separate circuit operated by a different voltage.

There are three factors that determine the type of control system: the power source the motor requires (voltage and phases), the conditions needed to start or stop the motor (pressure, temperature, etc.), and the current draw of the motor under normal conditions. Most single-phase motors will use single-pole or two-pole switches to operate them; 208/230 VAC single-phase motors often use the two-pole switches. Most three-phase motors, with a few rare exceptions, will use three pole switches. These three-pole switches will be contactors or motor starters, as we discussed in Lab 9.1. Three-phase motors are often large and expensive and will often have many safety controls. One of these is called a phase monitor, which can protect against phase loss, imbalance, or reversal, all of which can destroy a three-phase motor in a matter of seconds. Most large, expensive, or mission critical motors will contain a larger number of safety controls. You need to know what the purpose of a motor is in order to determine what operating condition controls it and which controls it may require. A compressor motor on an air conditioner, for example, will usually be switched on by a thermostat since the compressor should start when a building becomes too warm. A condensate pump motor might start when the level of water in the pan gets high enough so that when the pump motor starts it pumps the water out of the pan. There may also be some safety switches that are installed in the event of a problem.

PROCEDURE

Step 1. Using the shaded pole fan motor trainer, examine the wiring diagram or actual wiring and locate all the controls that will start or stop the motor; they should all be connected in series with the motor. List these below.

Step 2. Using the capacitor start motor trainer, examine the wiring diagram or actual wiring and locate all the controls that will start or stop the motor; they should all be connected in series with the motor. List these below.

Step 3. Using the three phase motor trainer, examine the wiring diagram or actual wiring and locate all the controls that will start or stop the motor; they should all be connected in series with the contactor or motor starter coil. List these below.

Step 4. Using the Figure 14-2-1 wiring diagram, create a legend for each control component. The legend should include a brief definition for each acronym and the purpose of the device. An example for the IBM is given. Use your text, the Electrical Diagrams packaged with your text, and instructor for assistance in finding out what each acronym represents.

Diagram Legend		Purpose of Device
IBM	Indoor Blower Motor	Circulates air over evaporator coil
BR		
F1, F2, F3		
CFM		
CC		
Comp		
LPS		
HPS		
FS		
TD		
PM		
F4		

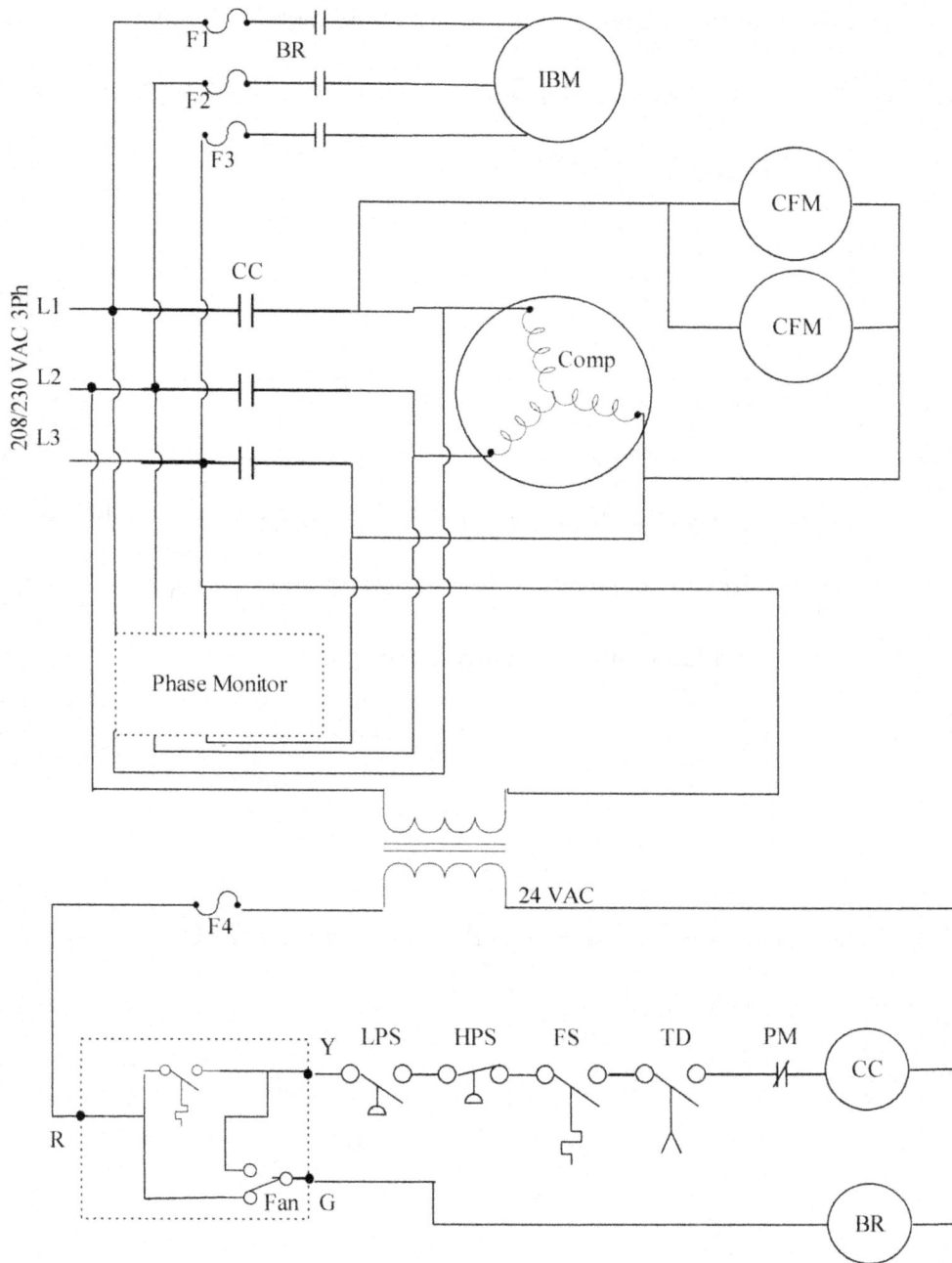

Figure 14-2-1. Use this diagram to fill in the table in Step 4.

QUESTIONS

1. Compare the controls you found operating the shaded pole motor in Step 1 with those you found operating the three-phase motor in Step 3. Were there more controls in one than there were in the other? Explain why you think this is the case.

2. As mentioned previously, certain types of motors require switches for operation or protection. What types of switches would you suggest for a refrigerant compressor motor? What type of switches might be needed for a fan motor? A pump motor?

3. In the wiring diagram used for Step 4, what is the phase monitor (PM) component used for? Is this a safety control or an operating control?

4. Using the diagram from Step 4, why do you think there are so many switches in series with the CC? Which motor is probably the largest (and most expensive to replace)?

LAB 14.3 Motor Troubleshooting

LABORATORY OBJECTIVE

In this lab you will learn how to quickly and effectively determine problems with all types of

motors and be able to determine if the problem is in the control system or with the motor itself.

ELECTRICITY FOR HVACR, 1e TEXT REFERENCE

Unit 14: How Motors Work

Unit 15: Motor Types

REQUIRED MATERIALS PROVIDED BY THE STUDENT

- DMM or VOM suitable for HVACR field work

- Capacitor tester (if required by your school)

- 6-in-1 screwdriver

- Various hand tools

REQUIRED MATERIALS PROVIDED BY THE SCHOOL

- Operational PSC fan motor

- Operational CSR compressor motor

- Operational three-phase motor

- Watt meter

- Various motors with winding problems

- Electrical tape

SAFETY REQUIREMENTS

This lab will involve applying high voltages to motors and capacitors. You should discharge all capacitors before handling them with an appropriate capacitor bleed resistor. Practice all lab safety precautions you have in the past when dealing with high voltages. This lab also demands special attention since we are working with motors that rotate very quickly. It is very easy to get a finger or clothing caught in a motor or fan blade. Be very careful when working around operating motors. Don't leave any tools or parts near an operating motor as they can be thrown out at a very high speed.

INTRODUCTION

If you were only able to take one measurement on a motor that was not operating correctly you would most likely want to measure current. Current draw is a very important measurement to take on a motor; it is a representation of how much work the motor is doing. There are two terms that you will need to understand to effectively troubleshoot motors: locked rotor amps (LRA) and full loads amps (FLA). Sometimes manufacturers will use the term rated load amps (RLA), which for the purposes of this lab is essentially the same as FLA. LRA is generally 5-8 times the FLA. Not all motors will have this information printed on them, but many will.

When a motor is operating and driving the fan blade, compressor, or pump as it is designed to it will draw FLA or very close to it. If a motor is trying to start and has a voltage applied to it but will not turn because of a seized bearing, stuck fan blade (or other driven component), or faulty starting component it will draw LRA. If you ever find a motor drawing LRA or pulling more current than the FLA your first step should be to determine if there is any problem with the circuit supplying power to the motor before condemning the motor. Each motor will have a specific voltage rating and it is important that the applied voltage be near this rating. Most manufacturers require that the voltage be within 10% of the voltage rating listed on the motor. That means if you have a 120 VAC motor the voltage supplied to it must be between 132 VAC and 108 VAC (10% of 120 is 12). If there is no voltage present then there is a problem with the control system providing power to the motor. If voltage is low then there is a problem with the power source. If there is proper voltage present you need to check the starting components the motor has. Troubleshooting starting components is essentially a process of elimination. Check the capacitors with a capacitor tester, but make sure you discharge it with an appropriate capacitor bleed resistor first. Testing relay's are a bit more difficult and will be covered in a later lab activity. Use the motor troubleshooting chart in Figure 14-3-1 to help you determine the cause of problems you encounter.

If you have determined that the starting components and voltage to the motor are all correct the problem is within the motor itself. The problem can be mechanical, like a failed bearing or stuck pump impeller, or it can be electrical. Electrical problems with a motor means that the windings of the motor are at fault. On all motors the windings can be simply checked with resistance. The motor will have a measurable resistance, which indicates a good winding; an infinite resistance indicates an open (and faulty) winding; and a very low resistance indicates a shorted winding. Another check that should be performed is a winding to ground test. With your DMM set for the highest resistance scale, check all motor connections to a ground point on the frame of the motor. Any measurable resistance indicates a grounded motor winding. Winding problems in all but the largest motors means that the motor needs to be replaced.

Note: Some motors have internal overloads and require that the motor is cool before the overload resets. If you measure an open winding in a motor with an internal overload make sure it is cool to the touch before determining the motor has failed.

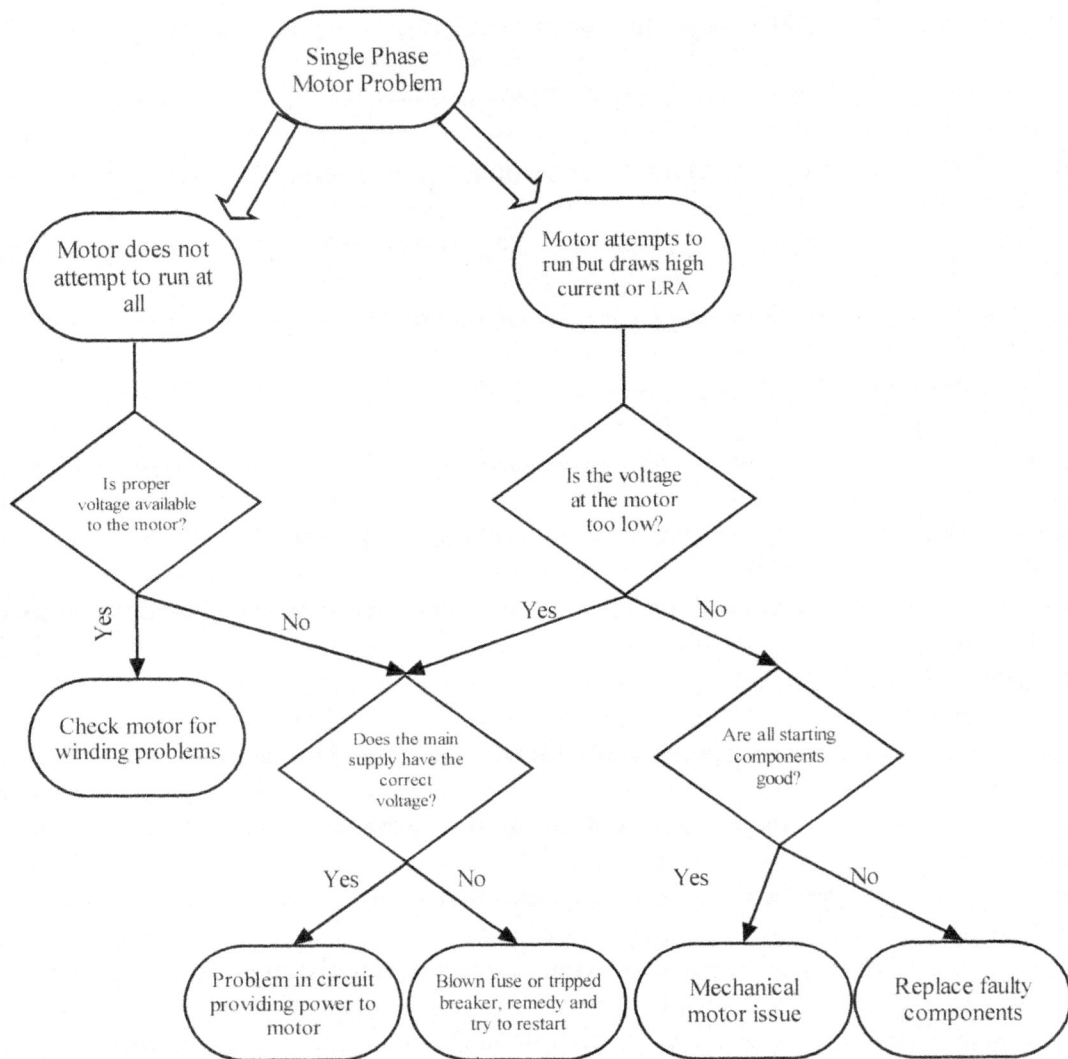

Figure 14-3-1. Use this troubleshooting chart to help you identify problems with single-phase motors. This is not an exhaustive list but will get you started in the right direction.

PROCEDURE

Step 1. Using the PSC motor trainer, start the motor and measure the running current; it should be near the FLA printed on the motor. Also measure the power consumption with the watt meter and record these values below.

_____ Amps

_____ Watts

Step 2. Once you know what current to expect, use a piece of cardboard or plywood to block off the airflow at either the return or supply air duct. Measure the current and power again and record these below.

_____ Amps

_____ Watts

Step 3. Using the CSR motor trainer, start the motor and observe the operation of the motor while running normally. You might want to take some voltage and current measurements while the motor is operating normally and record them below for reference later.

Step 4. With the CSR motor de-energized and the power source locked and tagged out, discharge the starting capacitor with an appropriate bleed resistor. Remove one terminal from the starting capacitor, wrap it with electrical tape, and lay it in a place where it will not get stuck in the motor or touch any metal framework.

Step 5. Install an ammeter on the motor and, after getting approval from your instructor, turn the power back on to the motor for a few seconds. Record the current you measure when the motor attempts to start.

SAFETY TIP: DO NOT LEAVE POWER ON FOR MORE THAN A FEW SECONDS!

_____ Amps

Step 6. After turning the power off to the motor and locking and tagging out the power supply, repair the capacitor connector. Remove the three wires on the compressor wiring terminals and measure the windings for proper resistance. Record your findings in the table.

Terminals to Check	Resistance Measurement (Ω)
Common - Run	
Common – Start	
Start - Run	
Common – Ground	
Run – Ground	
Start - Ground	

Step 7. Reconnect the wires to correct terminal on the compressor. After getting approval from your instructor, move to the next step.

Step 8. With the three-phase motor, measure the resistance of the windings in a similar manner to the way you checked windings in the CSR motor. Remember to follow all lockout/tagout procedures. Record your findings in the table.

Terminals to Check	Resistance Measurement (Ω)
T1 – T2	
T2 – T3	
T1 - T3	
T1 – Ground	
T2 – Ground	
T3 – Ground	

QUESTIONS

1. In the PSC motor you used for Steps 1 and 2 you should have measured a drop in amperage and wattage when you blocked off the airflow. Why would the motor draw less current and power when the discharge side is blocked? Do you think the same thing would happen with a positive displacement device like a compressor?

2. In Step 5 you measured a current close to or at LRA for the CSR motor. What problem is this simulating? What other conditions might cause LRA in this motor?

3. If the power was left on for a longer period of time in Step 5 what would eventually happen?

4. From the table you completed in Step 6, what do you notice about the resistance of the start winding (common – start measurement) compared to the run winding (common – run measurement)?

5. If you suspected a problem with the motor windings do you think you should check the winding resistances before or after applying power to the motor? What would probably happen if these windings were shorted or grounded and you applied power?

6. What is different about the winding resistance measurements you got for the three-phase motor in Step 8?

7. List below the possible causes of a motor that draws LRA.

8. You are working on a fan motor that just recently had a new fan blade installed. The original blade was not available and it was replaced with a similar size generic blade. The motor is drawing higher than FLA. No other changes were made to the system. What could be the issue?

9. Motors with internal overloads require special consideration when measuring winding resistance. What must you be sure of before condemning the motor?

Unit 15 Motor Types

Unit Summary

This unit provided you with much information about motors. Table 15-3 (page 279 of your textbook) provides a summary of the motors discussed. It includes the motor type, starting torque, running efficiency, operating cost, and usual applications. This motor comparison table will help you review and contrast the motors discussed in this unit. Finally, use Table 15-4 (pages 279-281 of your textbook) to aid in troubleshooting motor problems. Some of the conditions listed in the table were covered in this unit. This table will give you more ideas on what needs to be checked to solve problems. You will notice in the table that some of the problems are electrical and mechanical.

Key Terms (Definitions can be found in the Glossary in your text.)

Capacitor start/capacitor run (CSCR) motor

Capacitor start/induction run (CSR or CSIR) motor

Delta three-phase motor

Electronically commutated motor (ECM)

Permanent split capacitor (PSC) motor

Positive thermal coefficient (PTC)

Shaded pole motor

Single-phase motor

Three-phase motor

Wye three-phase motor

Unit 16 ECM: The Green Motor

Unit Summary

The ECM motor is found in high-end equipment. The OEM has many options for choosing the amount of airflow, on and off delays, and comfort profiles for each system. It is important for the installing contractor to set up the ECM operation based on the many motor options. Without a correct ECM setup, the system may not operate properly, meaning the customer will not be satisfied with the system's performance.

Current generation ECMs are much improved over earlier ones. Reliability improvements over the generations include the following:

1. Fully encapsulated module electronics to protect against moisture damage, the number one reason an ECM fails
2. EMI filters to provide protection against line transient and voltage spikes
3. Speed limiting to prevent overcurrent operation due to extremely high static pressure operation
4. Durable ball bearings on all models.

The ECM has more troubleshooting steps when compared to a conventional PSC motor. Even so, a high percentage of these motors are returned under warranty with no problem found. It is important for the ECM setup to be complete before beginning the troubleshooting process. This is especially true of a new installation. Check the input voltages. The motor is a DC, three-phase motor, therefore the motor windings should all have the same resistance. The resistances should be equal and be less than 20 ohms. Finally, an ECM test instrument is a valuable asset when diagnosing ECMs.

Work safely around an ECM. Disconnect the power to the ECM before removing the two connectors. Also wait about 5 minutes before separating the control module from the motor. Note that the large charged capacitors can hold a dangerous charge. Refer to Table 16-3 (pages 304-305 in your text) for help identifying the vast benefits of constant-torque and variable-speed ECMs when compared to PSC motors. It is good to know the benefits of the ECM so that you can talk intelligently with your customer about the features and benefits of this high-end product.

Key Terms (Definitions can be found in the Glossary in your text.)

Constant-torque motor

Electronically commutated
motor (ECM)

External static pressure (ESP)

Integrated control module
(ICM)

LAB 16.1 The ECM Motor

LABORATORY OBJECTIVE

The purpose of this lab is to familiarize you with a motor that is becoming more common in HVACR equipment: the electronically commutated motor, or ECM. As with any motor, it is important that you understand its operation, characteristics, and troubleshooting steps.

ELECTRICITY FOR HVACR, 1e TEXT REFERENCE

Unit 16: ECM: The Green Motor

REQUIRED MATERIALS PROVIDED BY THE STUDENT

- DMM or VOM suitable for HVACR field work

REQUIRED MATERIALS PROVIDED BY THE SCHOOL

- An ECM motor installed in a furnace or air handler

- A PSC motor installed in a furnace or air handler

- Watt meter

- ECM module tester

SAFETY REQUIREMENTS

This lab will involve working with motors and line voltage. ECM motors are no different than standard motors in their fast rotation. All previous motor safety precautions should be taken, including using caution when working around operating motors and when dealing with energized circuitry. Follow all lockout/tagout procedures when servicing the motor.

INTRODUCTION

ECM motors can come in two general varieties: variable speed and constant speed or torque. Variable speed ECMs are usually found in comfort heating and cooling air handlers. Constant speed/torque ECMs are usually found in refrigeration evaporators and certain propeller fans, as well as in some air handling units. Variable speed ECMs can be identified by the larger number of motor connections at the module and the programming dip switches on the circuit board. Constant speed/torque ECMs usually have two or three connections necessary to operate: line 1, line 2, and sometimes a ground connection. The electronics are always contained and sealed inside a constant speed/torque ECM.

ECM motors are essentially three-phase DC motors that are operated by an electronic controller. Since this controller is more complex than other motor starting components, manufacturer's provide ECM test modules for troubleshooting their ECM motors. These test instruments are a must for technicians and there a several types available. Certain ECM motors can be checked by measuring voltages at the connection pins, but some will require the use of the test instrument. The electronics on a variable speed, programmable ECM are found on the module connected to or near the motor and the circuit board that send the 24 VAC signal to the motor to start.

The motor section of an ECM is actually quite simple. They are three-phase DC motors that have three individual windings. We know from Lab 14.3 that three-phase motor windings should all be of equal resistance, usually less than 20 ohms each. Checking ECM motors involves checking the motor windings, checking the module and verifying proper supply voltage and connections. There are two sets of connections on variable speed ECMs: line voltage connections and low voltage connections. Most motor connections are made with a molded plastic plug and are difficult to connect incorrectly. One typical wiring problem that occurs is incorrect polarity. With some electronics, including ECM modules, polarity can cause incorrect operation. The first thing you should check when trying to locate a problem with an ECM is to check the power supply to the equipment for correct voltage and polarity. Motors we have dealt with in previous labs were not polarity sensitive and would operate normally even with the wrong polarity. The other voltage problem that may occur is the low voltage control circuit that signals the motor to start and stop when the thermostat opens and closes. The 24 VAC signal is usually present on the C terminal and one other, depending on if the thermostat is calling for heat or cooling. If the voltage is correct, the line voltage polarity is correct, and there is a correct 24 VAC signal, then the problem is likely with the ECM module or the three-phase motor itself. Use a module tester to determine if the module is good or not.

PROCEDURE

Step 1. Using the ECM system your instructor assigns you, examine the indoor blower motor and determine which type of ECM is installed, a variable speed or constant torque. Once you have examined the motor, start the equipment and observe the operation of the ECM, especially when it first starts.

Step 2. Shut down the system and perform lockout/tagout procedures.

Step 3. If this motor has dip switches on the circuit board to control speed, determine which switches are in which settings and verify the switches are in the correct position for this system.

Step 4. Disconnect the module from the motor and test for resistance at the three connections going directly to the motor. Remember to also check the windings to ground to check for a short to ground condition. Record your findings below.

Step 5. Reverse the polarity of the line voltage power supply at the motor by switching the L and N connection wires. If you cannot easily switch connections at the motor you can switch the polarity of the main power source to the equipment.

Step 6. Reconnect the module and replace all other connections the way you found them with the exception of the power supply wires you just changed. Restart the system and observe how the ECM operates with reversed polarity. Briefly describe your observations below.

Step 7. If the ECM is not functioning correctly, the first step is usually to verify correct supply voltage and polarity. Once that is verified testing the module is usually next. Using the module tester provided by your instructor connect it to the system as described in the instructions and check the ECM module.

Step 8. If you suspect that polarity is incorrect you must test it to verify. Usually the black wire is L or line 1, and white is neutral or line 2, but this is not always the case. With a 115 VAC supply measuring from line 1 to ground will give 115 VAC while the neutral (line 2) to ground will be at or near 0 VAC. With a 208/230 VAC power source both hot conductors to ground will measure 115 VAC. Verify which wire is line 1 and which is line 2 with your DMM.

Step 9. Correct the polarity of the wiring and then restart the blower motor and use a watt meter to measure the power consumption of the motor or entire system if necessary. Record your measurement below.

Step 10. Measure the power usage of a comparably sized PSC blower motor or system. If you measured the entire system power in Step 9 make sure you measure the entire system power for this step as well. Record your measurement here.

Step 11. With the system operating, block off the supply air duct with a piece of cardboard or similar object. What happens to the operation of the motor? Describe your observations in your own words below.

QUESTIONS

1. In Step 1 when you saw the motor start, did you notice anything unusual about how the motor started? What industry term describes this unique way of starting?

2. If the motor has dip switches on the board were they set correctly for this system? How would you need to adjust the settings if the furnace size was increased by 10,000 Btuh or the cooling was increased by ½ ton?

3. When you reversed the polarity in Step 5 the motor probably did not operate. How could this happen in the field (assume the system has been installed and operating normally for awhile)?

4. If the ECM you checked out operated for ten hours, how much energy in kWh would be used? How many kWh would be used for the PSC motor you measured if operated for the same ten hours?

5. Did you notice anything different about how an ECM shaft turns when you spin it by hand? Why does it do this?

6. In Step 7 you used an ECM module testing tool. Does this tester work for all ECMs? How does it indicate a good or faulty module?

7. What parts of an ECM can be replaced easily without having to replace the entire motor assembly?

8. In Step 11 you blocked off the supply air flow from the system. Why did the motor speed change? What would happen as the air filter in a system gradually loaded up with contaminants?

9. Using Table 16-1-1, indicate which dip switches should be in the "on" position for a 3.0 ton cooling system with normal air flow. If you measured the external static pressure and found it is 0.70 inches water column, how many CFM will this motor provide?

Outdoor Unit Size (Tons)	Airflow Setting	Dip Switch Setting						External Static Pressure				
		SW 1	SW2	SW3	SW4			0.1	0.3	0.5	0.7	0.9
2.5	LOW (350 CFM/TON)	OFF	ON	OFF	ON	CFM	WATTS	880 / 120	875 / 155	860 / 190	845 / 225	840 / 245
	NORMAL (400 CFM/TON)	OFF	ON	OFF	OFF	CFM	WATTS	1020 / 170	1000 / 205	990 / 240	980 / 280	960 / 320
	HIGH (450 CFM/TON)	OFF	ON	ON	OFF	CFM	WATTS	1110 / 210	110 / 260	1110 / 320	1100 / 350	1100 / 385
3.0	LOW (350 CFM/TON)	ON	OFF	OFF	ON	CFM	WATTS	1040 / 190	1010 / 220	1000 / 260	1000 / 310	990 / 340
	NORMAL (400 CFM/TON)	ON	OFF	OFF	OFF	CFM	WATTS	1200 / 250	1200 / 320	1190 / 370	1190 / 415	1175 / 450
	HIGH (450 CFM/TON)	ON	OFF	ON	OFF	CFM	WATTS	1340 / 355	1340 / 425	1330 / 475	1320 / 530	1300 / 570
3.5**	LOW (350 CFM/TON)	OFF	OFF	OFF	ON	CFM	WATTS	1215 / 265	1210 / 330	1210 / 375	1200 / 430	1185 / 465
	NORMAL** (400 CFM/TON)	OFF	OFF	OFF	OFF	CFM	WATTS	1430 / 415	1415 / 457	1410 / 520	1385 / 575	1330 / 580
	HIGH (450 CFM/TON)	OFF	OFF	ON	OFF	CFM	WATTS	1430 / 415	1415 / 475	1410 / 520	1385 / 575	1330 / 580

NOTES:
1. *First Letter may be "A" or "T"
2. **Factory setting
3. Continuous Fan Setting: Heating or Cooling airflow is approximately 50% of selected Cooling value.
4. For Variable Speed: low speed airflows are approximately 30% of listed values.
5. LOW 350 CFM/TON is recommended for Variable Speed application for Comfort & Humid Climate setting; Normal is 400 CFM/TON; High 450 CFM/TON is for Dry Climate setting

(Courtesy Trane Corporation.)

Table 16-1-1. Use this table to determine correct dip switch settings for the system in question 9.

Unit 17 Understanding Electrical Diagrams

Unit Summary

It is essential for technicians to understand electrical symbols and how to use a wiring diagram to troubleshoot. Compare the importance of this information to a truck driver who needs to know where to make a delivery. The driver must understand the symbols and roads shown on a map to get to the destination. A technician needs the same aids.

Understanding this information will help the reader in the upcoming troubleshooting units. You need to know the terminology in order to get help from the office or manufacturer's tech support hotlines. Unfortunately, the terminology is not always the same among technicians, so be aware of different terms in use in the field. The first part of this unit spent time defining terms such as circuit types: complete circuit, open circuit, short circuit, and grounded circuit. We also defined series, parallel, and series-parallel (combination) circuits. These circuit types are universally understood. The misunderstanding comes with the use of terms that relate to electrical diagrams. This unit provided what are considered to be the industry standard names for electrical circuits. Being able to draft a field diagram is a practical troubleshooting skill. There are many instances when the system wiring has been so badly butchered that it is a wonder the equipment is even operating. This can also be an unsafe condition. Electrical, pressure, and temperature safety devices are sometimes bypassed. Undocumented alternations make the wiring a rat's nest. This is observed when the tech opens the unit panel and a bundle of wire falls out. In such cases it is best to strip out all of the wires and rewire the system. A system-wiring diagram will be required for this step. Leave the new wiring diagram on the job and make a copy for future reference. Finally, practice drawing a wiring diagram and then converting it into a useful ladder diagram. When troubleshooting a diagram, use a clear page protector and erasable marker to trace out the circuit. Use different color markers to trace out different circuits. These steps will help you understand the various paths found in an HVACR circuit.

Key Terms (Definitions can be found in the Glossary in your text.)

Complete circuit	Pictorial diagram
Electrical diagram	Schematic diagrams
Ground circuit	Series circuit
Hopscotch troubleshooting	Series-parallel circuits
Ladder diagram	Short circuit
Line diagram	Symbols
Open circuit	Wiring diagrams
Parallel circuits	

LAB 17.1 Common Diagrams and Symbols

LABORATORY OBJECTIVE

In this lab you will identify common symbols used in HVACR Electrical Diagrams.

ELECTRICITY FOR HVACR, 1e TEXT REFERENCE

Unit 17: Understanding Electrical Diagrams

REQUIRED MATERIALS PROVIDED BY THE STUDENT

- 6-in-1 screw driver

REQUIRED MATERIALS PROVIDED BY THE SCHOOL

- Kit with 10 electrical components to identify

- Air cooled package system

SAFETY REQUIREMENTS

Make sure all power is disconnected and locked-out when identifying components. Keep hands safe from sharp objects and use common sense when working this lab. As always, wear safety glasses.

INTRODUCTION

In this lab you will begin to learn a different language, the language of electrical diagrams. This language starts with identification of the symbols. Most symbols are standard, but some can have many different applications. It is important to notice any letter designations that may be attached to symbol and always check the legend of a wiring diagram for the names. Although most symbols look like random drawings, they are actually designed to let you know what they do as well as how they work as you will find out when you begin to explore the symbols below.

PROCEDURE

Step 1. Identify the common electrical symbols below.

1. _____

2. _____

3. _____

4. _____

5. _____

6. _____

7. _____

8. _____

9. _____

10. _____

Step 2. Identify all the electrical components in the kit your instructor has put together. Write the name of the component, draw the symbol for the component, and give a brief explanation of the component's purpose (i.e., what it controls).

Component	Symbol	Letter designation	Component purpose

Step 3: Identify the components above in an air-cooled package system assigned to you by your instructor.

QUESTIONS

1. Explain the difference in a Heating thermostat and a Cooling thermostat.

2. What is bimetal and how is it used in electrical components?

3. Electrical symbols are shown in their normal position, which means they are energized or de-energized?

4. How can you tell the difference between a condenser fan and an evaporator fan in a wiring diagram?

5. Draw a single pole relay with one NO contact energized.

6. Dashed lines (---------) in a wiring diagram are an indication of what?

7. While troubleshooting a working unit, you read 0 volts across a set of contacts. Are the contacts open or closed?

8. What is the difference between a low-pressure switch and a high-pressure switch?

Unit 18 Resistors

Unit Summary

This unit discussed how to select a resistor based on a color code. Knowing a resistor's tolerance and power capabilities is important when purchasing a replacement component. Even though troubleshooting a resistor is not a common problem, it is important to be able to understand how to select a resistor. More and more systems are going to solid-state circuits, which use resistors, so it is important to know a little bit about a component we may see every day. There may be a time when you need to refer to technical material to determine resistance, especially if the resistor is open. Keep in mind that an electronic supply house will help you select an appropriate replacement.

Key Terms (Definitions can be found in the Glossary in your text.)

Carbon resistor

Resistor

Resistor tolerance

Wire-wound resistor

Unit 19 Fundamentals of Solid-State Circuits

Unit Summary

This unit was designed to give you a brief overview of electronics without overwhelming you with electron theory. We briefly discussed the operation of diodes, transistors, rectifiers, varistors, and circuit boards. Electron theory is important if you are working directly with electronic circuit boards. In our industry, however, circuit boards are good or bad. We do not need to know which component is defective. It is rare to attempt any field repairs on circuit boards, other than changing a blown fuse.

Key Terms (Definitions can be found in the Glossary in your text.)

Bridge rectifier	Microprocessor
Circuit board	Rectification
Diode	Rectifier
Integrated circuit (IC)	Semiconductor
Inverter	Solid-state circuits
Metal oxide varistor (MOV)	Transistor
Microcontroller	Varistor

LAB 19.1 Troubleshooting Printed Circuit Boards

LABORATORY OBJECTIVE

In this lab you will learn how to troubleshoot printed circuit boards by input voltage, output voltage, and the functions of a circuit board.

ELECTRICITY FOR HVACR, 1e TEXT REFERENCE

Unit 19: Fundamentals of Solid-State Circuits

REQUIRED MATERIALS PROVIDED BY THE STUDENT

- 6-in-1 screw driver

- DMM or VOM suitable for HVACR field work

REQUIRED MATERIALS PROVIDED BY THE SCHOOL

- Working system equipped with a multifunction printed circuit board.

SAFETY REQUIREMENTS

Follow all lab and shop safety rules, which includes wearing safety glasses at all times. Adhere to lockout/tagout procedures and never assume power is off, always check with a multimeter. It is unsafe to work around live electricity while wearing a watch or rings, so remove them before starting this task. Always make sure a low voltage fuse is in place to protect the transformer when checking low voltage power. When replacing printed circuit boards it is a safe practice to remove wire connections with insulated pliers. Many circuit boards have capacitors that can hold an electrical charge.

INTRODUCTION

Printed circuit boards have become very ordinary in HVACR equipment. A small circuit board can take the place of several relays and multiple safety switches, which can make troubleshooting much easier and lessens the number of components in HVACR equipment. There is a mystery about printed circuit boards because the average technician does not understand how or why they work. That is ok; in fact, this is one device that you are not expected to know everything about. You just need to know enough to determine if the device is good or bad. To determine this there are a few terms that you need to be familiar with: input voltage, output voltage, and function. The input voltage is just what the name implies, it is the voltage or electrical power going into the device. The electrical power can be low voltage, line voltage, or both. Output voltage is the electrical power leaving the circuit board to go do a task and could also be low voltage, line voltage, or both. The output voltage can come from many different terminals. Examples of output voltage could be electrical power leaving the circuit board to energize a fan motor, or energize a gas valve. Energizing these components is considered a function of the circuit board. Functions can also include monitoring safety devices like pressure switches or limit switches, as well as energizing components.

PROCEDURE

Step 1. The instructor will assign you a working system equipped with a circuit board. Find and record the model and serial number of the equipment, which is needed to order a replacement circuit board. Some circuit boards have identification on them, like the manufacturer's name and product numbers, if so record these also. Start the unit and record the input terminals and voltage and the output terminals and voltage of the board. The terminals are typically marked and/or labeled on the wiring diagram. From these voltage measurements you should be able to determine and record the many different functions of the circuit board.

Unit model number _____

Unit serial number _____

Circuit board identification information, if available:

Manufacturer's name _____

Model _____

Serial number of circuit board _____

Input terminals and voltage _____

Output terminals and voltage _____

List all the functions of the circuit board.

Step 2. Knowing the input voltage, output voltage, and functions of a circuit board is the main requirement for troubleshooting it. If you have voltage in on the input terminals and no voltage out on the output terminals, the board is bad and should be replaced.

From the circuit board functions above, list some of the many different problems you could possibly have that would justify replacing this circuit board. For instance, if there are output terminals to an indoor fan and a function of the circuit board is to energize the indoor fan, a possible problem could be that the circuit board failed to energize fan.

1. Function _____

Possible Trouble _____

2. Function _____

Possible Trouble _____

3. Function _____

Possible Trouble _____

4. Function _____

Possible Trouble _____

QUESTIONS: Answer the sentences below with terms from the text.

1. What is the term for electrical power entering into a printed circuit board?

2. Name the electronic device also known as an electrical check valve.

3. What is the name of a common one known as a MOV and used as a surge protector?

4. Name an electrical device that converts alternating current to direct current.

5. Name the device that took the place of the glass vacuum tube.

6. Name the solid state devices sometimes referred to as the chip.

7. What is the abbreviation for Light Emitting Diode?

8. What is the term for electrical power leaving a circuit board to energize components?

Unit 20 Taking the Mystery Out of Circuit Boards

Unit Summary

This unit provided an introduction to understanding circuit boards. You learned that circuit boards may seem complicated, but, with a little investigation, they can be understood. Even without the wiring diagram, several quick checks can help you determine if the problem is circuit board related or if the problem is with the inputs to the circuit board. For example, most circuit boards have low-voltage inputs and high-voltage inputs.

Measure the low-voltage input, which is usually 24 volts. The 24 volts can be supplied from a transformer that is located on the circuit board or supplied from an external transformer.

The high-voltage inputs can be 120 or 240 volts. The 120-V circuit is polarized, meaning that it is important to hook up the hot and neutral wires to the proper connections. A good ground is required for any voltage. Improperly wired hot/neutral voltages, or no or poor ground connections will make the circuit board operate erratically and sometimes not at all.

Check the circuit board's fuse. Other inputs you can check are pressure switches and other safety devices. Check labeled outputs on the circuit board. Again, these are checks you can do without the aid of a diagram or the flashing light error code table.

Having the wiring diagram and the flash code legend makes the troubleshooting process useful and logical. Use these resources before calling for help. Your supervisor or the manufacturer's tech support person will want to know what you have done. This is a professional practice. Record your troubleshooting findings on paper prior to making the call for assistance. This will help you understand what is happening and also assist your support person. With this documented information, tech support can make helpful suggestions.

Key Terms (Definitions can be found in the Glossary in your text.)

Black box	Low-pressure switch (LPS)
Circuit board	Reversing valve
Defrost thermostat	Solid state
High-pressure switch (HPS)	Test pins
Hot surface igniter (HSI)	Time-delay circuit

Unit 21 Air Conditioning Systems

Unit Summary

This unit covered air conditioning systems. A system is composed of many different parts and pieces that make up an operating air conditioning unit. This unit brought together the various bits and pieces that were presented in many of the preceding 20 units. We discussed single-phase and three-phase systems, package units, and split systems. We also covered a window unit diagram that had not been discussed up to this time. With the knowledge obtained in the prior units, you were able to understand an operating system.

This unit traced the sequence of operations of numerous air conditioning systems. The three-phase system discussed could also have been a refrigeration system used to cool products above freezing. It could not have been a freezer since the diagram did not have a defrost option.

Key Terms (Definitions can be found in the Glossary in your text.)

Bridge rectifier

Multispeed fan motors

Package unit

Split system

Systems

Window units

Unit 22 Gas Heating Systems

Unit Summary

This unit featured the types of gas heating systems and their operation. It started with a discussion about classifications of gas heating systems. It was noted that gas-burning systems could either be natural gas or liquefied petroleum (LP) systems. They can be further classified into efficiency types: low, medium, and high efficiency. These are known by their AFUE (annual fuel utilization efficiency) rating or percentage: 70% and below, 80%, and 90% and above. The unit then reviewed the operation of low- and high-efficiency systems. Low-efficiency systems have a simple design without many additional controls and components. This type of system has three electrical circuits: millivolts, low voltage or control voltage, and line voltage. The operation of the loads found in each of these circuits is controlled by related switches. The fan control operates the blower when the temperature of the heat exchanger is sensed, for instance. A safety control senses when too much heat is in the heat exchanger (high limit) and the thermocouple is used to determine if the standing pilot is lit. This provides safety protection. We do not want gas in the furnace if it cannot be safely ignited.

A high-efficiency system provides for flame and high-temperature safety, in addition to determining if the system is operating at its maximum efficiency by monitoring the venting of the exhaust gases. Exhaust gas pressure switches and electronic ignition help to conserve gas through more efficient gas burning. Because high-efficiency systems require more electrical and electronic devices, the technician has additional troubleshooting tools to use. The circuit board usually has an LED flash code that helps to verify observed problems in the operation. The technician may also have a troubleshooting flowchart to help guide the troubleshooting process. Whether it is a low- or high-efficiency system, the technician must make eight basic checks. These checks involve using test instruments to determine if the correct voltages are being applied. Some checks are visual or, in the case of temperature, conducted by a careful touch. As with any work conducted on gas heating systems, the work must be done using safe work practices consistent with industry policies and procedures. Always read and understand the manufacturer's installation and service manuals. Follow all safety procedures listed. Above all, use common sense. If you smell gas, do not operate light switches or create conditions that would cause ignition of those gases. A gas smell may be coming from a sewer system, but until it is investigated, all precautions must be taken to protect life and property.

Key Terms (Definitions can be found in the Glossary in your text.)

Electronic controls	Ignition system	National Fuel Gas Code
Flame sensor	Liquefied petroleum	Pressure switch
Gas control	Lockout	Type B venting
Gas heating	Manual reset flame rollout switch	Venting

LAB 22.1 Gas Heating Systems Controls and Wiring

LABORATORY OBJECTIVE

The purpose of this lab is to familiarize you with the operational sequences and common controls found in gas-fired heating systems.

ELECTRICITY FOR HVACR, 1e TEXT REFERENCE

Unit 22: Gas Heating Systems

REQUIRED MATERIALS PROVIDED BY THE STUDENT

- 6-in-1 screwdriver

REQUIRED MATERIALS PROVIDED BY THE SCHOOL

- Mid-efficient, induced-draft gas furnace

SAFETY REQUIREMENTS

Since you will be operating gas-fired equipment in this lab, caution should be taken to make sure the equipment is in proper operating order. The gas piping should be leak free, the flue gases should be properly vented, and all flammable and combustible materials should be removed from around the furnace. Check with your instructor to confirm that these safety precautions have been met. Many components of a furnace become extremely hot when burners are ignited so caution should be taken not to expose skin to hot surfaces. As usual, observe all lockout/tagout procedures, use common sense safety, and always wear safety glasses.

INTRODUCTION

This lab is to familiarize you with the operational sequence and safety controls of a gas furnace. Unlike air conditioning units, gas furnaces have many devices installed for safety reasons. If an air conditioning unit does not work, the homeowner is just hot and uncomfortable. If a gas furnace malfunctions without proper safety devices it can be a dangerous fire hazard.

It is very difficult for any technician to know all the components in every unit he or she will come in contact with, but if you can understand the sequence of operation of basic equipment it will make troubleshooting much easier. The sequence of operation is the order in which parts of equipment is energized, closed, opened, or ignited. If you know or understand the order components should be energized it gives you some direction on where to start troubleshooting. Most furnaces today, especially high-efficiency furnaces, are equipped with printed circuit control boards. These circuit boards monitor the many safety devices, control the indoor fan, and energize the gas valve. Some also have LED lights that will flash a fault code to inform the technician of a problem with the furnace. The fault codes are not meant to inform you of which part to replace, but to be used as a tool to let you know that there is a problem with the operation of certain devices and, when used with a troubleshooting flow chart, fault codes can be very helpful. You may not always have fault codes or troubleshooting flow charts to help, so it is important to start with the basics of identification, sequence of operation, and getting familiar with gas furnaces.

PROCEDURE

Step 1. Locate and identify the electrical components in an induced-draft gas furnace.

1. Gas valve

2. Hot surface igniter (NOTE: Do not touch igniter.)

3. High limit switch

4. Pressure switch or air proving switch

5. Roll out switch

6. Flame sensor

7. Inducer fan motor

8. Blower motor and list the number of motor speeds

9. Door switch

10. Transformer

Step 2. Locate and identify the gas heating components of a gas furnace

1. Burner

2. Gas manifold

3. Gas orifice

4. Heat exchangers

5. Flue vent

Step 3. Before turning on the furnace, perform a safety check. This includes making sure the gas piping is leak-free and the flue is vented properly; remove any combustible materials close to the furnace. Make sure to verify the correct incoming voltage to the furnace. Call for heat on the thermostat on the induced-draft gas furnace. List the operational sequence in the table.

Induced Draft Furnace Operating Sequence	
Order	Action
1	Thermostat set to heat
2	
3	
4	
5	
6	Thermostat set to OFF
7	
8	
9	

Step 4. The questions below are to be answered while inspecting a working mid-efficient furnace with inducer fan assembly. Always make sure electrical power is off when unplugging or unwiring devises. Use insulated pliers when removing wires from printed circuit board. Care should be taken with bare wires to avoid shock and the possibility of and electrical arc. The operation of the furnace is what the furnace is doing or where it is in the sequence of operation.

1. With furnace power off, disconnect the hot surface igniter being careful not to touch igniter. Turn furnace power on, call for heat. What is the operation of the furnace?

2. With furnace power off, disconnect air proving switch. Turn furnace power on, call for heat. What is the operation of the furnace?

3. With furnace power off, disconnect indoor blower wires. Turn furnace power on, call for heat. What is the operation of the furnace?

4. With furnace power off, disconnect **W** thermostat wire from furnace. Turn furnace power on and thermostat on call for heat. What is the operation of the furnace?

5. With furnace off, disconnect the flame sensor, turn furnace on heat. Turn furnace power on and thermostat call for heat. What is the operation of the furnace?

QUESTIONS

1. What is the purpose of a door switch?

2. What is the purpose of an inducer fan?

3. What is the name of the component that determines if flame is present after the burners ignite?

4. Name the electrical device that will turn burners off if the blower does not move enough air through heat exchangers.

5. Why is it important not to touch a hot surface igniter?

6. Name the component that senses the pilot flame on a low-efficiency furnace.

Unit 23 Electric Heating Systems

Unit Summary

Electric heating systems are smaller and less complex than fuel-burning systems that deliver the same number of Btu's as larger fuel-burning systems. Because they do not convert fuel to electricity, they do not need venting systems. With no need for venting systems, they can be placed anywhere and cost less to install. Because of their flexibility, electric heat systems can be produced to fit nearly any heating application.

Electric heating systems can be primary sources of heat or secondary sources. They can be installed as an individual room heating system or as an electric furnace to heat an entire building as a central heating system. Often an electrical heating system is coupled to a heat pump to provide backup heat during defrost periods or as supplemental heat during periods of low outside temperature.

How "green" are electric heat systems? Several ways of looking at electrical heating systems were discussed in regard to being energy efficient or using renewable energy. Electric heat is clean, convenient, and flexible. But if electricity can be produced by renewable sources, would direct electrical heat be considered a "green" technology?

The unit concluded with a small section on troubleshooting. Many of the troubleshooting techniques used for transformers, relays, and safety devices are also used for electric heating systems. Electric systems tend to be simpler in both operation and service than fuel-burning units. There are also fewer components to check during a service call. Some things that are checked on other types of furnaces become simpler to check on electric systems. Airflow, for instance, is a simple check. With four measurements and a few basic calculations using a formula, a technician can easily determine the airflow and adjust the fan speed.

Key Terms (Definitions can be found in the Glossary in your text.)

Adjustable thermal overload	Electric heat
Baseboard heaters	Fusible link
British thermal unit (Btu)	Gate controlled
Electric backup heat	Sequencers
Electric furnace	Silicone controlled rectifier (SCR)

LAB 23.1 Electric Heating System Controls and Wiring

LABORATORY OBJECTIVE

This lab is meant to familiarize you with the controls and operation of an electric heating system.

ELECTRICITY FOR HVACR, 1e TEXT REFERENCE

Unit 23: Electric Heating Systems

REQUIRED MATERIALS PROVIDED BY THE STUDENT

- 6-in-1 screw driver

- DMM suitable for HVAC work

- Ammeter

REQUIRED MATERIALS PROVIDED BY THE SCHOOL

- Thermometer

- Working electric furnace

SAFETY REQUIREMENTS

Follow all lab and shop safety rules, which includes wearing safety glasses at all times. Adhere to lockout/tagout procedures and never assume power is off, always check with a multimeter. It is unsafe to work around live electricity while wearing a watch or rings, so remove them before starting this task. Care should be taken when working close to electric heaters when power is on because they will have voltage to them. Even though a heating element is not energized, many heaters will feed power to one side (120 VAC) of the heater elements at all times. The heater will energize or heat up when power is applied to both sides (240 VAC) of the element. Common sense practices are a must in all lab situations, if something does not look safe, it probably is not safe.

INTRODUCTION

This lab is designed to get familiarize you with the controls and operation of an electrical furnace. You will locate and identify the main components that are responsible for controlling and monitoring electric heating elements.

Electric heat is used in many residential and commercial applications, whether it is for the main source of heat or as a secondary heat source for a heat pump. It can be costly to heat a home in a cold weather climate using just electric heat, which is why some states have changed their energy codes so that no new home can be constructed with an electric furnace as a stand-alone appliance. It can be used as a secondary heat source, but not the primary.

Electric heat is clean burning, is very flexible in the location of the appliance, and does not require venting. Another advantage of total electric heat is that it is easy to check the airflow of the system because you do not have to factor in fuel conversion. In Step 2 below you will have the opportunity to perform a CFM (cubic feet per meter) measurement. Knowing how to measure airflow can be very beneficial in servicing not only heating systems but air conditioning systems, too. Low airflow can result in a heating element getting too hot and burning out, so it is important that you do not block airflow while servicing an electric heater.

PROCEDURE

Step 1. Locate and identify the different components of the electric furnace you have been assigned.

1. Transformer

2. Fusible link

3. Low voltage fuse

4. Heating element

5. Thermal overload

6. Sequencer

7. Indoor blower; note the number of speeds

8. Ceramic insulators

Step 2. Your instructor will assign you an electric furnace to perform a CFM measurement. Allow enough time for all heating elements to energize; remember that heating elements are usually staged with time delay relays. Follow the instructions below to perform the task.

- Measure and record in the provided bank the amp draw of the operating heating element (or elements), including the blower amperage. All system heating elements should be energized for this measurement. This can be done at the unit disconnect.

- Measure and record in the provided bank the operating voltage.

- Measure and record in the provided bank the return air temperature.

- Measure and record in the provided bank the supply air temperature. This air temperature must be measured at three diameters of the plenum (at least), downstream of the heating elements. The temperature probe must also be shielded from the radiant energy given off from the elements (if there is direct line of sight). The airflow must be mixed to get an accurate temperature measurement. One way to ensure the accuracy of the temperature measurement is to place the temperature probe past the first supply air elbow so the heat coming directly off the heat strips will be mixed.

Total amperage, including blower _____

Operating voltage _____

Return air temperature _____

Supply air temperature _____

Use the following formula to complete the measurement:

Cubic feet per minute (CFM) = $\dfrac{\text{voltage (V)} \times \text{amperage (A)} \times 3.4 \text{ (Btu's per watt)}}{\text{supply temperature} - \text{return temperature} \times 1.08}$

CFM _____

QUESTIONS

1. What does Δ T mean?

2. What is the recommended CFM per ton of refrigeration?

3. Can you be shocked by touching an electric heater element?

4. Name the two common safety devices in an electric heater.

5. What is the amperage of a 5 KW heater operating at 240 VAC?

6. What is the simplest overload protection device?

Unit 24 Heat Pump Heating Systems

Unit Summary

In this unit we discussed the wiring and electrical diagram for a generic heat pump. We note that there are common components in every heat pump:

1. System control (thermostat, microprocessor, etc.)
2. Compressor and a starting system
3. System reversing valve
4. Safety shutdown controls
5. Fluid system motors (pumps/blowers/fans)
6. Emergency/backup heat.

The operation of a generic heat pump was described as it would operate in the cooling mode, heating mode, and defrost mode. When in the heating mode, an air-source system would need to go into defrost mode to shed moisture accumulation on the outside coil. A geothermal system would not need a defrost mode, but would need to monitor the temperature of the ground loop with a freeze sensor.

Troubleshooting was described in a general way. Remember that the manufacturer's installation manual should be consulted for an expanded troubleshooting table. The manufacturer may also have instructions on putting the microprocessor into diagnostic mode to aid the technician in solving the service problem.

Key Terms (Definitions can be found in the Glossary in your text.)

Air-source heat pump (ASHP)

Auxiliary heat

Defrost control

Demand defrost

Emergency heat

Energy Star

Geothermal-source heat pump (GSHP)

Ground loop

Groundwater system

Heat pump

Loss of charge pressure switch

Reversing valve

Test pins

LAB 24.1 Heat Pump Controls and Wiring

LABORATORY OBJECTIVE

In this lab you will become familiar with heat pump controls.

ELECTRICITY FOR HVACR, 1e TEXT REFERENCE

Unit 24: Heat Pump Heating Systems

REQUIRED MATERIALS PROVIDED BY THE STUDENT

- 6-in-1 screwdriver

REQUIRED MATERIALS PROVIDED BY THE SCHOOL

- Air to air heat pump with service manual

SAFETY REQUIREMENTS

Follow all lab and shop safety rules, which includes wearing safety glasses at all times. Adhere to lockout/tagout procedures and never assume power is off, always check with a multimeter. It is unsafe to work around live electricity while wearing a watch or rings, so remove them before starting this task. Care should be taken working around refrigerant lines because some may be hot. Common sense practices are a must in all lab situations; if something does not look safe, it probably is not safe.

INTRODUCTION

In this lab you will be introduced to the defrost circuit board, which is the main electrical control of a heat pump. Most modern heat pumps use printed circuit boards to control defrost because of the many functions a circuit board can do and the small size of the component. Older units would use a series of relays, timers, and switches to perform the same task as one circuit board. The defrost circuit board is not the only way a heat pump can defrost, but is has become the most common.

What makes a unit a heat pump is the reversing valve that changes direction of the refrigeration cycle. In the cooling mode, a heat pump absorbs heat at the evaporator coil inside and rejects or gets rid of that heat outside at the condenser coil. That is why an air conditioning condensing unit blows warm out the top in the summer time. In the heating mode a heat pump absorbs heat from outside and rejects that heat inside to warm a space. The air blowing out of the condensing unit is now cold and, if the ambient temperature is low, the condensing unit will ice over and freeze. When the condensing coil is frozen, it cannot absorb heat, so to keep the heating cycle productive, a defrost cycle is necessary.

The defrost board monitors and controls many functions during this defrost cycle. The first is to determine when defrost is needed by monitoring a defrost thermostat (DFT) at specific time intervals. If defrost is needed, the defrost board shifts the reversing valve to the air conditioning mode and de-energizes the condenser fan so that the outdoor coil will warm and melt the ice. The defrost board also sends low voltage inside to energize the auxiliary heat because we are now in the air conditioning mode and, without some form of heat, the homeowner will have cold air blowing inside. When the defrost cycle is complete, the unit will resume normal heat operation. The defrost thermostat typically is the device that starts the cycle by closing when the coil temperature gets low enough and ends the cycle when the coil temperature gets warm enough.

PROCEDURE

Step 1. On the heat pump you have been assigned, remove the control panel and locate the defrost circuit board.

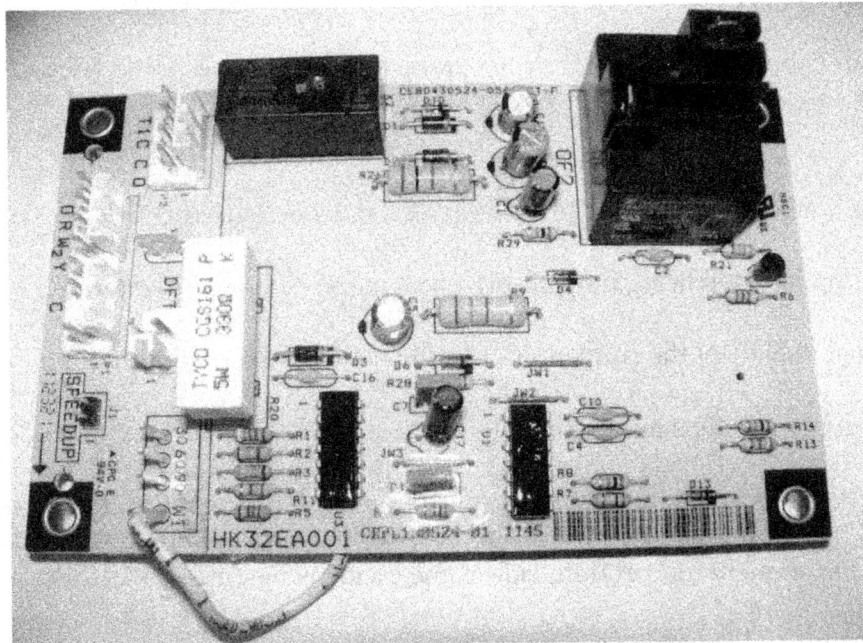

Figure 24-1-1. A common heat pump defrost circuit board. Pins at the bottom left are the time setting; above the time setting pins are the defrost thermostat connections marked DFT. The outdoor fan relay is in the top right corner.

Record the time setting of the unit you have been assigned.

Time setting of the board _____

Locate the wiring to the reversing valve, defrost thermostat, and outdoor fan. It is important to understand that the defrost circuit board controls the operation of these components.

Step 2. Locate the defrost thermostat on the heat pump your instructor has assigned you. You may have to follow the wires from the defrost board. Once the thermostat has been located, disassemble the unit so you can remove the defrost thermostat. Record the temperature at which it closes. The temperature is usually listed or stamped on the thermostat.

Defrost thermostat temperature _____

Step 3. Turn the power on to the heat pump you have been assigned. Set the thermostat to cool, call for cool, and wait for heat pump to come on. There is usually a three-minute delay. While the heat pump is running, disconnect the thermostat wire to the reversing valve, which is typically the orange wire. If the reversing valve shifts, then this unit energizes the reversing valve in the cooling mode. If you do not hear the reversing valve shift, turn the thermostat mode to off and wait for the indoor fan to turn off. Now run the heat pump in the heating mode; call for heat. While the unit is running in the heating mode, disconnect the reversing valve wire and listen for the reversing valve shift. If you hear the reversing valve shift, this unit energizes the reversing valve in the heating mode.

When is the reversing valve energized on this unit?

Circle one: heating mode   cooling mode

A typical heat pump thermostat will have an O terminal for energizing reversing valve in cooling mode and a B terminal for energizing reversing valve in the heating mode.

Step 4. On the heat pump you have been assigned, you will do a defrost circuit board test. Troubleshooting a defrost board can be very difficult because of the time setting on the board and the temperature of the DFT, so unit manufacturers commonly have a sequence of procedures you can do to check the operation of the defrost circuit board. The defrost board test procedures are typically on the unit service panel or in the manufacturer's service procedure information. Ask your instructor for help locating this information and proceed with the test. Have your instructor initial lab when test has been run.

QUESTIONS

1. Why is W2 wired to the defrost board on a heat pump?

2. Why is the condenser fan de-energized during the defrost mode?

3. What is the component responsible for changing the refrigeration cycle from cooling to heating cycle?

4. The terminals the defrost thermostat is connected to are labeled as what?

5. A heat pump thermostat typically has two different terminals that can be used for wiring the reversing valve. How are the terminals labeled and what is the purpose of the two?

6. When a heat pump is defrosting, is the reversing valve in the cooling mode or heating mode?

7. When the outdoor temperature is low and humidity is high, would you need to increase the defrost time setting or decrease the defrost time setting?

Unit 25 How to Start Electrical Troubleshooting

Unit Summary

Electrical troubleshooting is the process of identifying and eliminating problems with the ultimate result of finding and correcting the fault. The troubleshooting process is very individual and there are many avenues to obtaining the final result of getting a system back on line. This unit outlined some recommended methods to first do a quick review of the job and see what is working and what is not working. Some problems can be very simple, whereas other problems can be very complex.

Each technician must develop a comfortable troubleshooting technique. The ACT method was recommended as a way of organizing a step-by-step sequence for troubleshooting. The ACT method gives the technician a way to conduct a quick check of the major components while getting familiar with the system under inspection.

This unit discussed many effective and ineffective ways of troubleshooting. The hopscotch method was explained as one of the better troubleshooting tools. The technician was encouraged to call or seek help for more difficult problems. Prior to seeking help, the technician should document important operating and model information on the system.

Key Terms (Definitions can be found in the Glossary in your text.)

ACT

Hopscotch troubleshooting

Line side

Schematic diagram

Troubleshooting

LAB 25.1 Using a DMM for Basic Troubleshooting

LABORATORY OBJECTIVE

The purpose of this lab is to familiarize you with basic troubleshooting pictures you will encounter in the field. The goal is to make you competent in measuring voltage, current, and resistance, and to know when each measurement is appropriate.

ELECTRICITY FOR HVACR, 1e TEXT REFERENCE

Unit 25: How to Start Electrical Troubleshooting

Unit 26: Basic Troubleshooting Techniques

REQUIRED MATERIALS PROVIDED BY THE STUDENT

- DMM or VOM suitable for HVACR field work

- Various hand tools

REQUIRED MATERIALS PROVIDED BY THE SCHOOL

- A heating and cooling system with a fault

- A heating and/or cooling system with a burned out transformer

- An operational heating and cooling system

SAFETY REQUIREMENTS

Voltage and current measurements must be taken with power applied to a circuit, so we will be working with "live" circuits and components. Do not plug in the circuit until you have received the approval of your instructor. Remember to follow safe hand-tool use practices and make all electrical connections neat and orderly to prevent wires from accidentally touching. Lock out/tag out procedures should also be followed. You should review Unit 3 in your textbook before attempting this lab assignment to make sure you know how to operate your meter when taking voltage and current measurements.

INTRODUCTION

Successful electrical troubleshooting begins with taking appropriate measurements at the right time. Voltage, current, and resistance are all useful values to know under the right conditions. The key is to not only know when to take each measurement, but also what to expect when you do measure a value. If you don't know what to expect the measurement is essentially meaningless. If you were to measure air pressure in a tire and find it is 50 psi, it would not have any useful value if you didn't know the pressure was supposed to be 35 psi. When you have a reference you can make a judgment and say the pressure is too high.

This lab will focus on checking individual components only and will provide you with some tips on how and when to take these measurements. Let's start with a familiar light bulb. How do you verify it is burnt out? A light bulb is a type of load and all loads have some measureable resistance. You can remove the bulb and check for the appropriate resistance. The other option is to check the fixture base for voltage. If the correct voltage is available to a load but it is not functioning, the load has likely failed. Something a little more challenging might be a relay or contactor. We know that these are made up of two parts: the coil and the contacts. Since the load is a coil it should have some measureable resistance. We could also check operation of the coil by measuring voltage at the coil when it is connected to the circuit. If the correct coil voltage is present and the relay contacts are not changing position then the relay has failed. The contacts are more challenging because to verify operation of the contacts the coil needs to be powered, therefore voltage is the best option. The other method that could be used is applying a coil voltage while checking for open or closed switches at the contacts. Remember, there cannot be any voltage present when you are measuring resistance with your DMM, so make sure you measure the contacts and not accidentally try to measure resistance across an energized coil. Voltage is usually the first check you will perform to see if the load should be operating. Voltage is also the method of choice when checking through switches. Resistance is usually the second measurement you take when you have identified an open switch or faulty load with your previous voltage checks.

Current is also a useful troubleshooting tool. Current can tell you that a load is operating and, in the case of motors, how much of a load it is under. Let's say you were wondering if an electric heater is functioning in a furnace. A quick current check would verify that it is if it's pulling current and operating. This is a quicker method than shutting off power, disconnecting the heater, and measuring resistance since a correct current measurement indicates the heater is good. An open heater or other load will not draw any current and you would need to investigate further. Remember, with loads that pull greater than 10 or 20 amps (depending on your meter) you need to use a clamp-on ammeter. Measuring current through motors is especially useful as we discussed in Lab 14.3. Some loads, like relay coils, draw very low current, in the milliamps (mA) or microamps (μA) range. If you do not have a meter that measures low current accurately you can use a loop of wire wrapped 10 times and connected in series with the load. If you clamp your ammeter around all 10 loops and take a measurement the reading will be multiplied by 10. That means if your meter shows 1.5 Amps AC (AAC), the actual current is 0.15 AAC. Figure 25-1-1 demonstrates this concept.

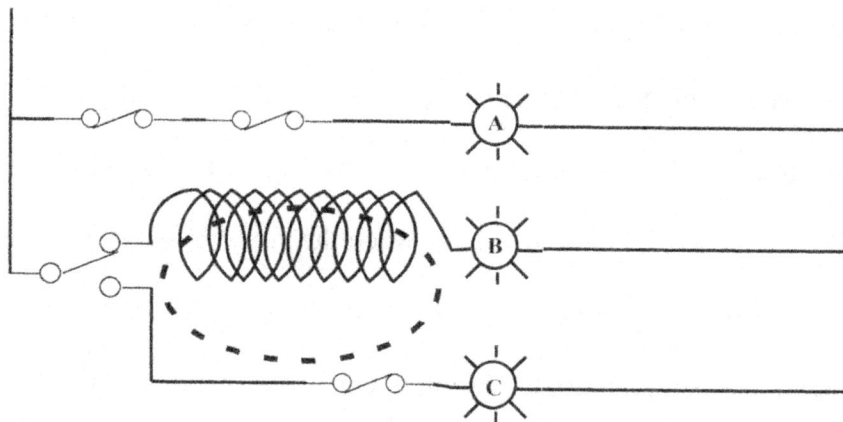

Figure 25-1-1. This figure shows a 10-wrap coil of wire installed in series with load "B." This amplifies the reading you will get with your clamp-on ammeter, simply divide the reading by 10 to get the true current.

When faced with a blown fuse or tripped circuit breaker you need to find why the device opened first before you replace a fuse or reset the circuit breaker and leave the job. Since these devices open based on high current you should take a current reading when trying to restart the system. Most of the fuses and circuit breakers are time-delay types in HVACR systems. If the breaker trips or fuse blows immediately, there is a short circuit or short to ground, which will be covered in Lab 27.1. If the fuse blows or circuit breaker trips after a short time delay there is an overload problem, most likely a motor is not starting.

PROCEDURE

Step 1. Using the "bugged" heating and cooling unit that your instructor assigns, locate an electrical diagram for it and then determine which load is not functional in the system. Examine the system thoroughly and use your senses to collect information about the system.

Step 2. Your first thought might be that this load is faulty. Your next step should be to verify the correct voltage at the load. Carefully measure the voltage across the load and record the value below. Make sure you record what to expect before measuring the voltage.

Expected VAC _____   Measured VAC_____

Step 3. If there is correct voltage present across the load then you have determined the fault is with the load and you can skip Step 4 and proceed to Step 6. If the correct voltage across the load is not present you need to check the switches that control that particular load.

Step 4. Switches can be checked with resistance when the power is off or voltage when the power is on. If the switch does not open or change switch position upon de-energizing the circuit you can check it with resistance. If you are trying to check the switch of a relay or similar switch, use voltage drop to determine if the switch is closed or open. Test all switches in the control circuit using voltage or resistance to see if they are open or closed. Record your measurements below; make sure you know what to expect before you take a measurement. Normally this trial-and-error method of troubleshooting is not recommended, but it is used here to illustrate checking for voltage drops across individual switches. If you have more than three switches to test in the circuit add additional spaces as needed.

Switch #1   Expected reading_____   Measured reading_____

Switch #2   Expected reading_____   Measured reading_____

Switch #3   Expected reading_____   Measured reading_____

Step 5. Now that you have found an open switch, what caused the switch to open? Determine when this switch opens and then check that condition. Did this switch fail or did it open for a good reason? Write your explanation below.

Step 6. Using the heating and cooling system with a burnt out transformer, try turning the system on and observe what happens. Which loads do not function? Which do?

Step 7. Which measurements would you need to take to confirm that the transformer is faulty? Take the appropriate measurements that verify the transformer is faulty.

Step 8. Transformers do not simply "wear out," they usually fail because of a short in the secondary 24 VAC circuit, which may be a wire touching a ground or a faulty relay coil. The overload current can be small, around 1.6 amps for a 40 VA transformer, and some clamp-on meters will have difficulty measuring this value accurately. Using the operational system assigned by your instructor, install a 10-loop wire in series with the secondary side of the transformer "R" wire. Use your clamp-on ammeter to measure the current; make sure you clamp around all 10 wraps. Record the value you measure below.

_____ Indicated Amps   _____ True Amps (reading divided by 10)

Step 9. With the operational system, block the indoor blower motor with a piece of wood or other object your instructor suggests. Hook up a clamp-on ammeter to one of the line voltage leads for the motor and start the system. Turn the power on for only as long as you need to get a good current reading. Record your current measurement below.

Step 10. Return the system to normal operation and put all tools and equipment back in their correct locations.

QUESTIONS

1. When faced with a load like a motor, heater, or coil that is not operating, which measurement is usually performed first?

2. Which check is usually performed second?

3. What could current checks be used for?

4. When an automatic switch is found to be open the first step is never to replace it. What do you need to do first?

5. For the system with the faulty transformer, it probably showed similar symptoms to having a line voltage source not providing power. How do you think you could tell if it is a line voltage problem or low voltage problem?

6. From your answer in Step 8, is this transformer's secondary coil overloaded? Use the following formula to determine maximum current the transformer secondary coil can provide.

$$\text{Maximum current of secondary} = \frac{\text{VA of transformer}}{\text{Secondary Voltage}}$$

7. In Step 9 you measured a higher than normal current when the blower motor was seized. Was it enough to trip the breaker or blow the fuse in the service disconnect? What other loads in this system could cause an over-current condition that would trip a breaker or blow a fuse?

Unit 26 Basic Troubleshooting Techniques

Unit Summary

When troubleshooting, define the problem, locate the problem, and fix it. Prior to troubleshooting ask yourself the following questions:
- What should the unit be doing?
- What is the unit doing?
- From this you will be able to determine: What is the unit *not* doing?

Knowing what is supposed to happen in an HVACR circuit is important to the troubleshooting process. Unfortunately, knowing exactly how the circuit is supposed to operate is not always possible.

It is important to realize that most problems are not difficult to solve. By following the steps in this unit, a technician can quickly check many of the common problems found in HVACR systems. Even the most experienced tech requires help from time to time. First, try to solve the problem without help through the use of the logical troubleshooting sequence you have chosen. Your "logical sequence" may be different from others' logical sequence. You need to select a system that is right for you and one that will help you figure things out.

Key Terms (Definitions can be found in the Glossary in your text.)

ACT

Clamp-on ammeter

Control voltage

High-pressure switch

Lockout

Low-pressure switch (LPS)

Sequencer

LAB 26.1 Electrical Troubleshooting – The Hopscotch Method

LABORATORY OBJECTIVE

In this lab activity you will be learning to use the hopscotch troubleshooting method discussed in your textbook to find breaks or open switches in several circuits. Hopscotching is a very smooth and reliable process used to find open switches, broken wires, and faults within a circuit. The goal is for you to be able to pick out the correct circuit to troubleshoot and then to use a ladder diagram to find the test points for your leads in the circuit, then move one meter lead through the circuit to find the problem.

ELECTRICITY FOR HVACR, 1e TEXT REFERENCE

Unit 25: How to Start Electrical Troubleshooting

Unit 26: Basic Troubleshooting Techniques

REQUIRED MATERIALS PROVIDED BY THE STUDENT

- DMM or VOM suitable for HVACR field work

REQUIRED MATERIALS PROVIDED BY THE SCHOOL

- A basic wiring trainer for each student or group

- Several faulty components to insert into the training board

SAFETY REQUIREMENTS

Voltage and current measurements must be taken with power applied to a circuit, so we will be working with "live" circuits and components. Do not plug in the circuit until you have received the approval of your instructor. Remember to follow safe hand-tool use practices and make all electrical connections neat and orderly to prevent wires from accidentally touching. Lock out/tag out procedures should also be followed. You should review Unit 3 in your textbook before attempting this lab assignment to make sure you know how to operate your meter when taking voltage and current measurements.

INTRODUCTION

Electrical troubleshooting is best learned in a hands-on environment as it takes a great amount of practice to master troubleshooting skills. Once you understand the basic methods we will cover in this lab you will find learning how to troubleshoot more complex circuits is easy. Electrical problems are usually black and white, meaning the circuit or component is working or it is not. Occasionally you will encounter an intermittent problem, which can be frustrating and time consuming, we will cover some tips about those in the advanced troubleshooting labs. When trying to determine whether the problem is due to an electrical fault or not, you simply need to understand the system and how it functions. Take a forced air furnace, for example. You should know that when there is a call for heat from the thermostat the draft inducer, blower, and gas valve should be powered up an operating. If one or more is not (either working intermittently or not at all), then you can assume the problem is electrical. If everything is functioning, then the problem is not electrical in nature. Occasionally you will encounter an electrical problem where a load does not turn off when it is supposed to. These are rare, but can occur when a switch or contact is stuck closed.

Because most HVACR equipment uses relays to operate it is important that you learn to troubleshoot with voltage. Many technicians might try to use resistance to check switches and circuits because they feel more confident in it and it is safer since there is no power applied. The problem is that many circuits change function when there is no power; an electronic thermostat when powered down will reading open resistance to the cooling or heating terminal even if the thermostat is working properly. The only way to verify if the contacts are closing is to power it up, and then we cannot measure resistance anymore and depend on voltage. One popular method of voltage troubleshooting is called *hopscotching*. It is named this because one meter lead moves through the circuit in a hopscotching fashion while the other remains stationary. You can find breaks in a circuit by finding voltage differences throughout a circuit.

There are two ways of hopscotching: the stationary lead on the L1 side of the circuit and the stationary lead on the L2 or "load" side of the circuit. The L2 stationary lead is used more frequently since it is more intuitive and easier to make sense of. In Figure 26-1-1 you can find an example of hopscotching a simple circuit that has an open switch "TC." Notice how voltage is indicated with the moving lead all the way up to the point of the TC, but when the lead crosses over the voltage drops to 0. This is the indication that voltage is not getting through this switch and it is open.

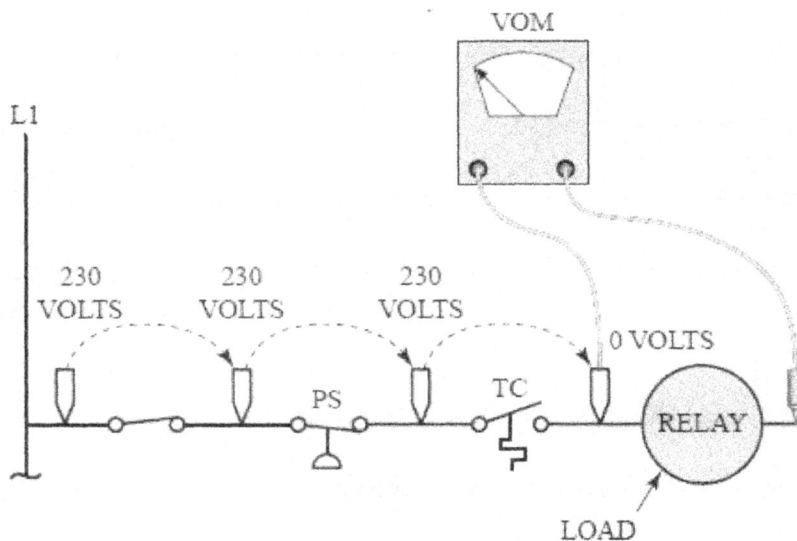

Figure 26-1-1. This demonstrates the basic hopscotching process. Notice how there is voltage through all the switches up to the last switch labeled TC. This diagram has an open TC. This circuit is not functioning; if the TC switch was closed it would energize the relay coil and work properly.

If you were hopscotching this circuit and measured 230 VAC on the left side of "TC" and 0 VAC on the right, you can determine that the TC is open. It might be open because it failed or it might be open because it is sensing a high temperature condition and it is supposed to be open. Also notice how we were looking for a <u>difference</u> in voltage, not always 0 VAC. Some circuits will not show 0 VAC across an open switch. You should realize that, as you travel through the circuit, any difference in voltage indicates a problem. Voltages should be exactly the same on both sides of a closed switch.

Helpful Hopscotching Hints:

- It is important to make sure the circuit you are testing is <u>supposed</u> to be operating while you are troubleshooting it.

- Make sure you know where your leads are in the ladder diagram, otherwise you don't know what your testing.

- Don't pay any attention to the circuits that are working. If a compressor is running and the indoor blower motor is not, what sense is there in troubleshooting components that control the compressor?

- You should know what reading you will get before you take a measurement, if you just measure 120 VAC it won't help unless you knew what to expect.

- If you don't know how a piece of equipment is supposed to operate you cannot effectively hopscotch it, find out first.

PROCEDURE

Step 1. Use Figure 26-1-2 to determine where the open switch in the circuit is. Circle the component in the diagram that is open.

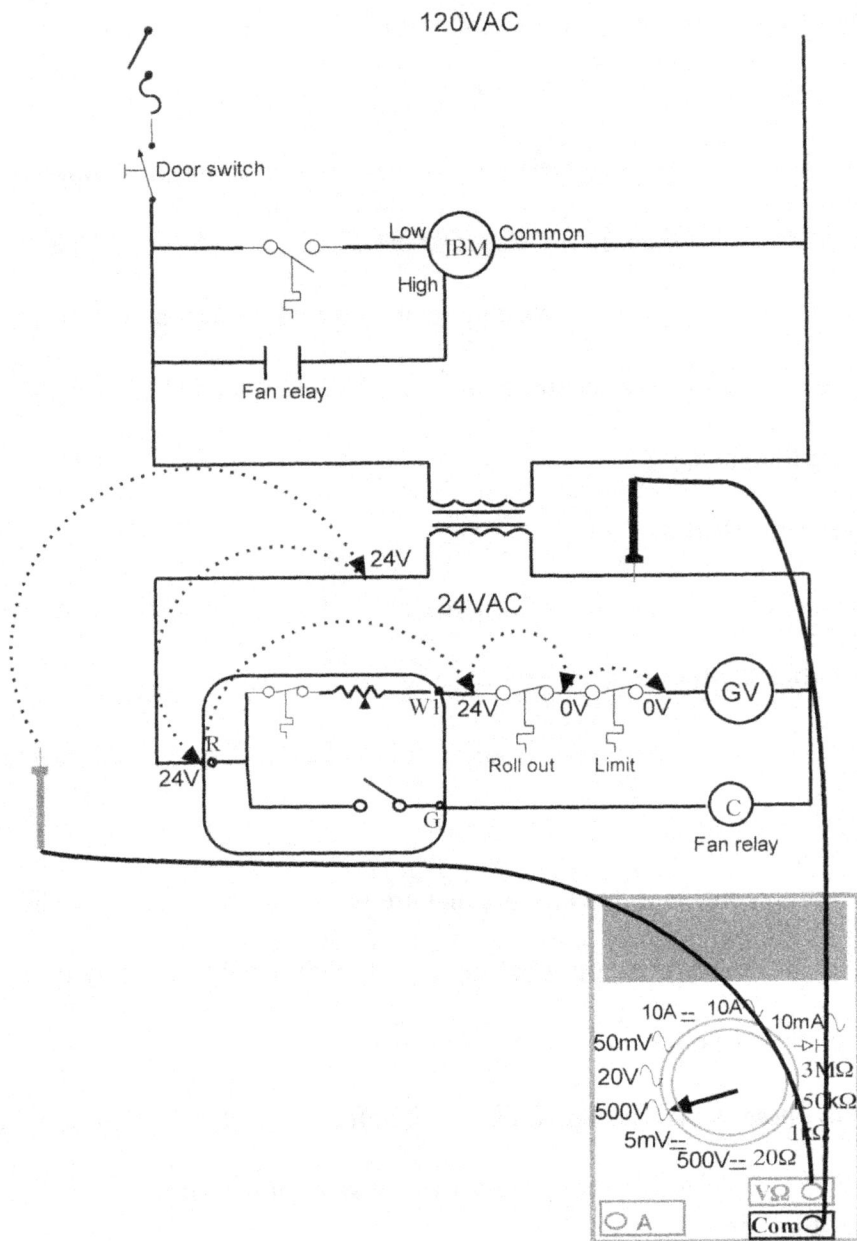

Figure 26-1-2. This diagram is used for the first step in the procedures.

Step 2. Use Figure 26-1-3 to identify the break in the circuit. Circle the problem in the diagram.

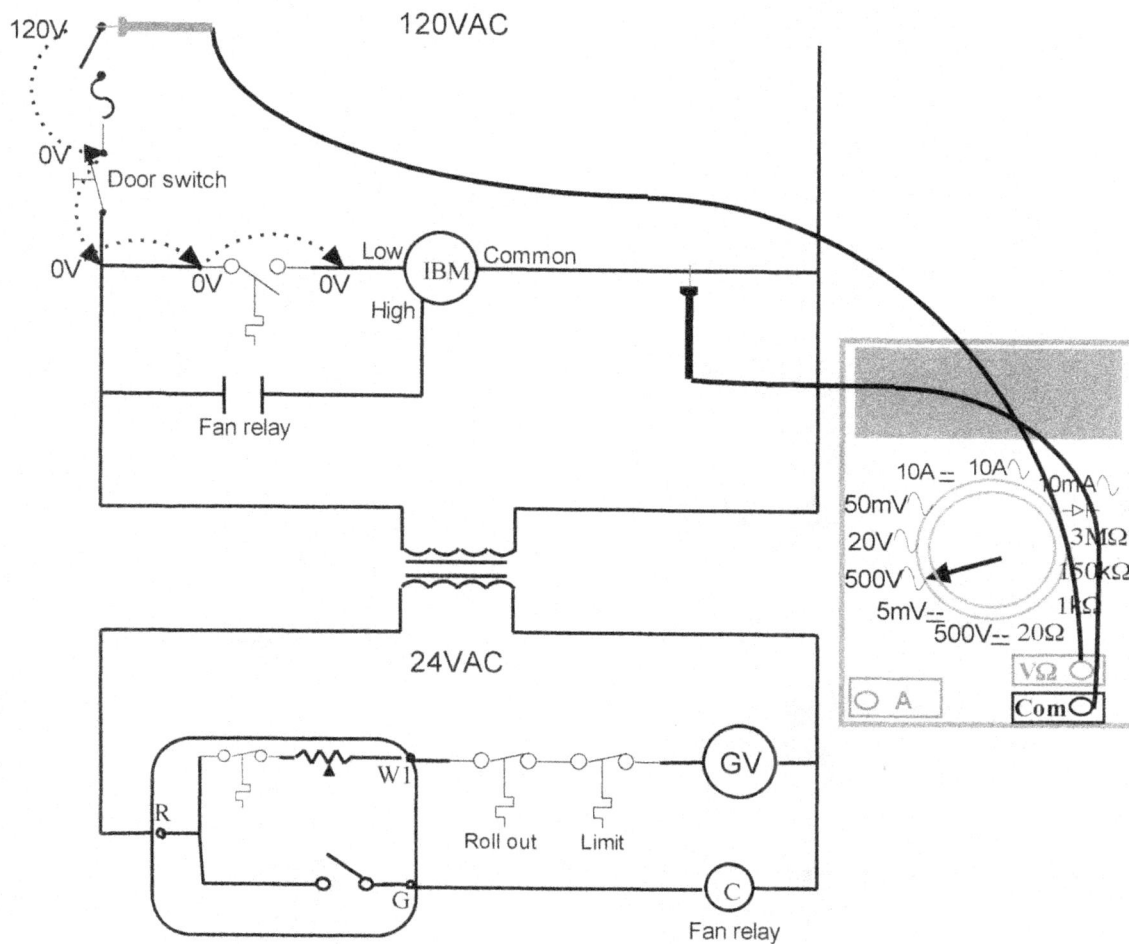

Figure 26-1-3. This diagram is used for the second step in the procedures.

Step 3. Use Figure 26-1-4 to determine the problem with this circuit. Write your response

below.

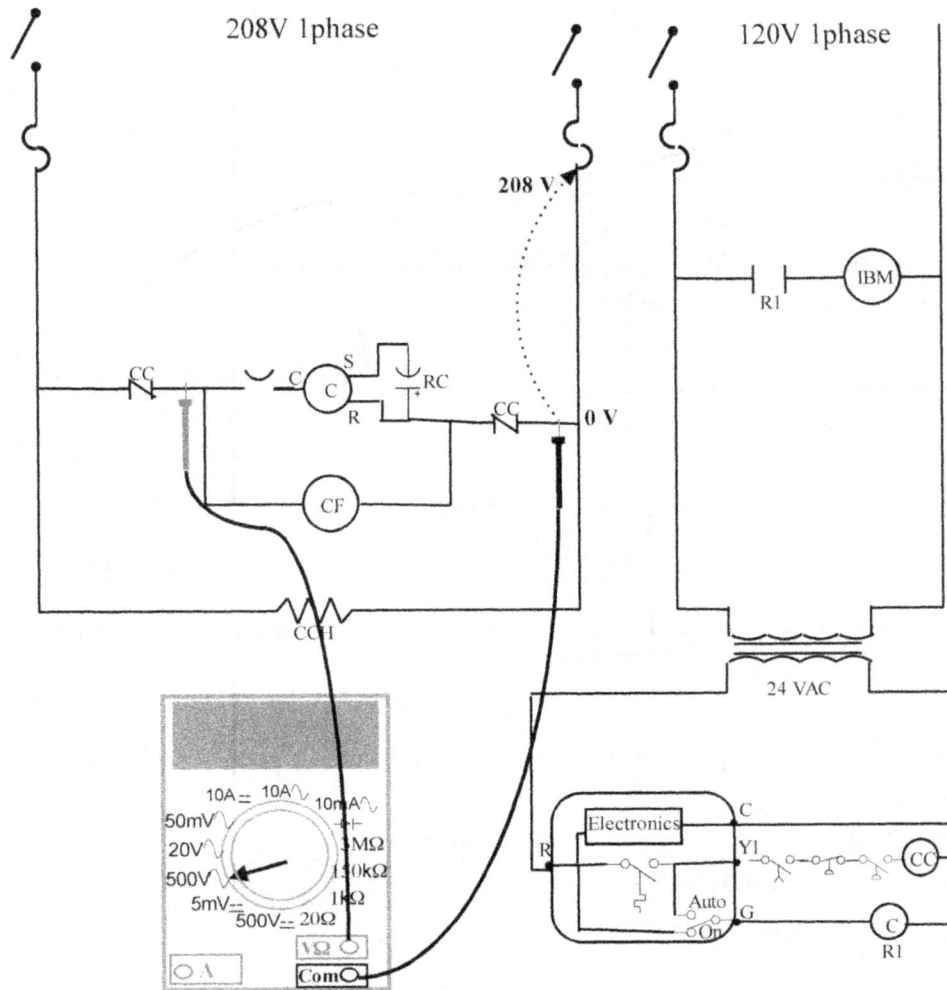

Figure 26-1-4. This diagram is used for the third step in the procedures.

Step 4. Using the basic wiring trainer, wire the circuit found in Figure 26-1-5 into the trainer. Make sure it works correctly when you are finished.

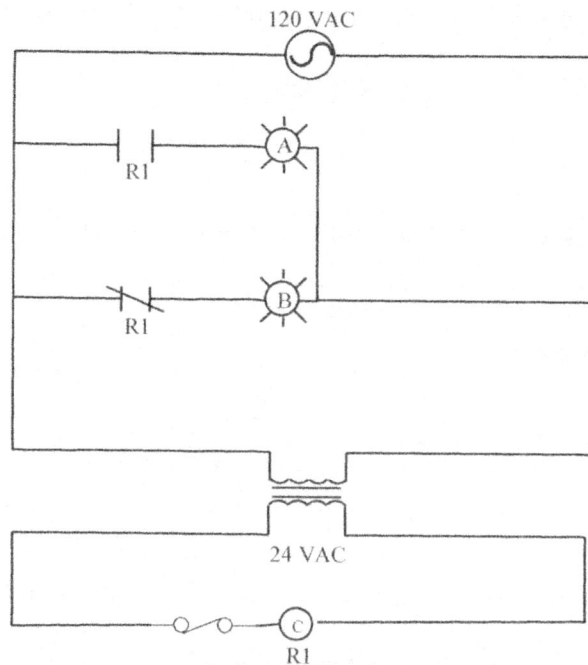

Figure 26-1-5. This diagram is used for Step 4 in the procedures to wire into your basic trainer.

Step 5. Using a faulty component provided by your instructor, replace a good component in your circuit with the faulty one you selected. Examples of "bugs" to install might be replacing the relay with one that has an open coil or stuck contacts, replacing the transformer with a defective one, inserting a wire that is broken internally, replacing the SPST switch with one that does not operate, or replacing one of the light bulbs with a burnt out one. Get approval from your instructor before moving on to the next step.

Step 6. Switch trainers with one of your classmates or another group, don't tell them the problem you put in the trainer. Using the other person's/group's circuit, use your DMM to find the fault they put in. Do not start replacing or try to randomly look for the problem. Find out which loads are not functioning correctly and then use your DMM to hopscotch the correct circuit. Write the fault you found below.

Step 7. After locating the problem verify your predictions with the group you traded off with. Return all tools and parts to their correct locations.

QUESTIONS

1. The best measurement to check when troubleshooting is voltage in the circuit when power is applied. Why?

2. In Step 1 you found an open roll out switch. Is the solution going to be to replace the switch with another? What other steps should be taken when any open switch is identified?

3. In Step 2 you were troubleshooting a blower motor circuit for a heating system. When you first arrived on this job what do you think you would have noticed to lead you to checking this specific circuit?

4. The fault you found in Step 3 wasn't an open switch, what would you need to do to fix the problem?

5. You hopscotch a circuit such as the one in Figure 26-1-4 and find there is correct voltage through the line side (L1) switches all the way up to the load. When you take your two probes and measure voltage directly on either side of the motor you find there is 0 VAC. If there is no problem in the line voltage going through the switches where must this problem be?

6. When you measure proper voltage at a load but it is not operating as it is supposed to, what do you do next?

7. Once you determine a component is faulty using voltage, how can you verify your conclusion?

Unit 27 Advanced Troubleshooting

Unit Summary

The purpose of this unit was to give you troubleshooting practice for various types of equipment. We spent a considerable amount of time discussing compressor troubleshooting.

The unit applies some of the skills that you have learned in other sections of this book. After talking with the customer, it is important to do a quick system survey using the ACT method or some logical sequence for determining what is operating. There is really no one good way to troubleshoot. The sequence you choose for troubleshooting will differ from that of other techs. Every tech has different ways of approaching problems or "figuring things out." The ultimate goal of troubleshooting is to find the problem and solution in a reasonable amount of time. As you do troubleshooting you will develop ways to speed up the process, create your own shortcuts, and pass your tips on to newer techs entering our profession. You will learn from others as they learn from you. We are in this together to help each other become better technicians.

Key Terms (Definitions can be found in the Glossary in your text.)

ACT

Digital multimeter (DMM)

Hard start kit

Locked rotor amps (LRA)

LAB 27.1 Advanced Troubleshooting

LABORATORY OBJECTIVE

In this lab you will learn how to perform troubleshooting on more complex components and systems. You will learn the techniques used to find open circuits, shorted conditions, and other problems that can cause higher than normal current. You will also be able to select universal replacement parts from a catalog if the OEM parts are not available.

ELECTRICITY FOR HVACR, 1e TEXT REFERENCE

Unit 27: Advanced Troubleshooting

Unit 28: Practical Troubleshooting

REQUIRED MATERIALS PROVIDED BY THE STUDENT

- DMM or VOM suitable for HVACR field work

- Capacitor tester (if required by your school)

- Various hand tools

REQUIRED MATERIALS PROVIDED BY THE SCHOOL

- An operational split-cooling system

- An operational three-phase system

- A cooling system with a low charge

- A small 2 or 3 mA glass body fuse and fuse holder

- Compressor starting box or cord

- A cooling system with a faulty compressor starting component

- Compressor service handbooks

- Parts catalog or other source for compressor starting components

SAFETY REQUIREMENTS

Voltage and current measurements must be taken with power applied to a circuit, so we will be working with "live" circuits and components. Do not plug in the circuit until you have received the approval of your instructor. Remember to follow safe hand-tool use practices and make all electrical connections neat and orderly to prevent wires from accidentally touching. Lockout/tagout procedures should also be followed.

INTRODUCTION

Troubleshooting involves using every tool you have at your disposal. In addition to your tools, you will find manufacturer's wiring diagrams, compressor service literature, installation manuals, and other manufacturer notes useful. As mentioned in Lab 25.1, you should have some expectation of what you will find when a measurement is taken. These pieces of literature will give that information to you. In today's electronic world access to the information you need is often a phone call or website away, take advantage of the technology you have.

There are certain electrical problems in systems that can be difficult to find. Among them are intermittent electrical problems. When you are servicing a system that fails occasionally the information your customer provides is very important. They hopefully can give you the information you need to find the problem. You might want to find out the specific time of day or how often the failure occurs. When the failure occurs, ask them to tell you exactly what happens; does the indoor blower continue running or does everything completely shut down? Does a circuit breaker trip? These answers can lead you to the problem much more efficiently than guessing. Once you know the load(s) that are not operating you are left with looking at the controls that operate that load. It is likely a switch that is opening. A trick to identify the switch that opens occasionally is to install a very low current fuse, usually a few milliamps, in parallel with the switch you think is opening. When the unit fails again, current will be forced through the fuse and it will blow since it is not rated for a very high current. When you return, if the fuse is blown you know that this is the switch that opened, if not it was a different switch.

Short circuits can also be frustrating, especially if fuses are being blown every time the system is turned on. The circuit breaker or fuse is not always at the equipment so when you reset a breaker it might trip quickly before you can get back to the equipment to see what the problem was. It helps to have a coworker with you to help identify short circuits. Sometimes you can check major load resistances with an ohmmeter to find the problem. Otherwise, if the short or short to ground is not immediately obvious by arcing or burn marks, the best method is to disconnect and isolate suspected loads or circuits and then, with an ammeter on the circuit, turn the power back on and see if the current still exceeds normal levels. If it does then the problem is in another device. If current returns to normal levels the component or circuit you disconnected contains the short circuit. This will help save you from creating a pile of blown fuses unnecessarily.

PROCEDURE

Step 1. Using an operational split air conditioner that your instructor assigns, locate the following information, which could be used for troubleshooting the system if it had a problem. Use any resource you have available to you; not all of the information will be found in the manufacturer's literature.

Outdoor Unit Model Number	
Outdoor Unit Serial Number	
Indoor Unit Model Number	
Indoor Unit Serial Number	
Compressor FLA	
Compressor LRA	
Compressor Motor Type (PSC, CSIR etc.)	
Condenser Fan FLA	
Condenser Fan Type (Shaded Pole or PSC)	
Outdoor Unit Voltage Rating	
Indoor Blower Motor FLA	
Indoor Blower Motor Type (Shaded Pole or PSC)	
Indoor Unit Voltage Rating	
Refrigerant Type	
Safety and Operating Switches Used in System	

Step 2. With the three-phase system provided by your instructor, measure for correct current balance and voltage balance. Refer to page 531-533 in your text for information on how to check for these imbalances. Record the three indicated % imbalance values below.

% supply voltage imbalance _____

% supply current imbalance _____

% load current imbalance _____

Formulas:

$$\% \text{ imbalance} = \frac{\text{maximum deviation from average voltage}}{\text{average voltage}} \times 100$$

$$\% \text{current imbalance} = \frac{\text{maximum deviation from average current}}{\text{average current}}$$

Step 3. Using the cooling system with a known low charge, install a gauge set and verify the conditions it is operating at. The system should be cutting off on the low-pressure switch. If the low-pressure switch was only opening intermittently and you could not catch it happening you would want to install a fuse across the switch in parallel with it. When you return on a repeat failure you could see the fuse was blown and know this was the switch responsible. Turn off the system and install a small 2 or 3 mA fuse in parallel with the low-pressure control. Get approval from your instructor and then restart the system and observe the fuse blowing when the low-pressure switch cuts out. You now would know that the low-pressure switch is the one that has been opening.

Step 4. Using the system with the faulty compressor assigned by your instructor, start up the system and assess the operation of all the loads. The compressor should not be operating while the rest of the system is. The first thing you want to check is to verify there are no short or short to ground conditions in the motor windings. Lock out/tagout the power source and check the compressor start and run windings for correct resistance according to manufacturer specifications. Make sure no short or ground conditions exist. The compressor electrical service handbooks may list the resistance each winding is supposed to be. Record your findings in the provided table. Keep in mind that the compressor windings will appear open if the compressor has an internal overload and it has tripped. If you find an open winding give the compressor time to cool down before re-taking your measurements.

Common to Run Resistance	
Common to Start Resistance	
Run to Start Resistance	
Common to Ground Resistance	
Start to Ground Resistance	
Run to Ground Resistance	

Step 5. If the windings check correctly, the second thing you will want to do is verify there is voltage at the compressor common and run terminals. Record your measurement below.

VAC from Common to Run_____

Step 6. If the voltage is correct, then take a current measurement on your compressor when it tries to start. If there is no voltage, then you need to check the operation of the contactor and control circuit. Compare the measured current to manufacturer's specifications. Record your findings in the table below.

	Manufacturer's Specifications	**Your Current Measurement**
Compressor FLA		
Compressor LRA		

Step 7. Your compressor is likely drawing much more than the specified FLA and is close to LRA indicating the compressor is not turning. The problem could be a mechanical problem or a starting component problem. The best way to tell is to remove all starting components from the compressor and install a starting box on the compressor. If the compressor successfully starts with the manual stating box and runs normally for a few seconds you have a starting component

problem; if it does not start and continues to draw LRA then the compressor has failed mechanically. Make sure you have a clamp-on ammeter installed around the common wire from the start box when you try to start the compressor to measure current. Do not let the compressor run for very long.

Step 8. Now that you know the issue is with a starting component, test the capacitors with the capacitor tester to make sure they are within specifications. If they check out, then you most likely have a starting relay problem (if there is one). Lockout/tagout the power supply and then use the table below to compare the resistance readings of the relay's type you have with what they should be.

Relay Type	Correct Resistance Readings	Your Measured Resistance
Potential Relay	Terminals 1-2 = 0 Ohms (contacts) Terminals 2-5 = 1kΩ-5kΩ (coil)	
Current Relay	Terminals L-M = less than 10Ω (coil) Terminals S-L or 2= ∞Ω (contacts)	

Step 9. If these all check out then the relay contacts are probably stuck in the closed position (potential relay) or open position (current relay). To test for this, wire the compressor back up as you found it and turn the power back on. If you have a potential relay, right after the compressor tries to start, pull off the #2 terminal wire on the relay, which disconnects the start winding manually; if the compressor starts and runs normally the relay was stuck closed. If you have a current relay it is a bit more difficult to test since the relay is often mounted directly on the compressor terminal pins. The switch is opened by gravity so when the relay is upright in the

mounted position the contacts should be open and when you turn it upside down the contacts should close. But if you have ruled out all other options, the likely candidate is the relay.

Step 10. Using any installed system your instructor assigns, look up a hard start kit that contains a relay and starting capacitor. Use the parts catalogs or any other source your instructor recommends. List the part number below.

Hard Start Kit Part No. _____

QUESTIONS

1. You are working on a cooling system that has obviously burnt compressor starting components. You do not have the correct components and notify the customer they would have to be ordered. Do you think there is anything else you should check before you order replacement parts?

2. From Step 2, were all the phase imbalances you calculated within the 2% maximum voltage and 10% maximum current imbalance guidelines? If they were not, what would you do to correct the problem?

3. Where might you find useful literature for troubleshooting a system other than what is available on the unit?

4. In this lab you used a compressor starting box to manually start a compressor. What components does this starting box replace?

5. In Step 10 you selected a hard start kit for an air conditioning system. Why might you need to add one?

6. If you are troubleshooting a system and find the wiring diagram had been lost or destroyed, what are your best options when trying to hopscotch the system?

7. Look at ED-10 included in your electrical diagram package. This system is an indoor heating section of a system with the cooling system not shown. There are a number of electrical diagram notes the manufacturer has provided. Indicate why you think they may have included the following notes:

Note 5: Lines inside PCB are printed circuit board conductors and are not included in legend.

Note 6: Replace only with a 3-amp fuse.

Note 10: Ignition lockout will occur after four consecutive unsuccessful trials for ignition. Control will auto-reset after three hours.

Unit 28 Practical Troubleshooting

Unit Summary

The goal of this unit was to help you practice electrical troubleshooting using wiring diagrams and to help you learn in-depth troubleshooting of some basic components found in HVACR systems. This unit presented some troubleshooting problems, provided their solutions, and described how to use a logical sequence to solve the problems.

HVACR problems were presented along with several ways to solve the problem. There may be many different ways to solve a problem, but usually there is one best way to derive a correct solution. This unit was not so much about finding the most correct way of solving a problem, but about solving it in a reasonable amount of time. Solving the problem in a reasonable amount of time includes calling for help when advice is needed instead of wasting time on the job. Everyone needs help sometimes. Your troubleshooting skills will improve with experience. Learning how to troubleshoot is the most important skill you will need as a service technician. Lesser skilled techs can change out a contactor or compressor. But determining that those components are defective takes more knowledge, skill, and experience. Troubleshooting skills come with practice and listening. Listen and learn from other techs who have found solutions to HVACR problems. It seems that one common trait of HVACR techs is their willingness to share a success story. You can learn from others' experiences without being on the job; of course, the best learning tool is figuring out the problem yourself.

Use your "lifelines" when you need help. Do not call for advice in front of your customer. Sometimes it is a good idea to step away from the problem for a short while. Go on a short break or get gas for your vehicle. Step back and evaluate the information.

Key Terms (Definitions can be found in the Glossary in your text.)

Hopscotch troubleshooting

Ladder diagram

Appendix A: Common Symbols Found in HVACR Diagrams

CAPACITOR

* NEAREST GROUND

CIRCUIT BREAKER

3-POLE CIRCUIT BREAKER WITH THERMAL OVERLOAD DEVICE IN ALL 3 POLES.

GROUND

COIL OPERATING

COIL OR RELAY CONTACTOR

* ADD LETTER DESIGNATION

MECHANICAL CONNNECTION

(SHORT DASHES)

MECHANICAL CONN./FULCRUM

(SHORT DASHES)

CLOSED CONTACT

RELAY   SWITCH

OPEN CONTACT

RELAY   SWITCH

TRANSFER

RELAY

SWITCH

OPEN RELAY CONTACT

WITH TIME CLOSING

TC

CLOSED RELAY CONTACT

WITH TIME OPENING

TO

FUSE

CARTRIDGE OR PLUG FUSE

FUSIBLE LINK

MOTOR (GENERAL)

MOTOR WINDINGS

MAIN

AUX.

SINGLE PHASE

(TYPICAL)

CONDUCTORS

FACTORY POWER
FACTORY CONTROL
FIELD POWER
FIELD CONTROL

CONDUCTORS (CROSSING)

CROSSING OF PATHS OR CONDUCTORS NOT CONNECTED.
(ANY ANGLE)

CONDUCTORS (CONNECTED)

RESISTOR

(HEATING)

SWITCHES (TYPICAL)

THROW SWITCHES

SINGLE THROW

DOUBLE THROW

DOUBLE THROW
DOUBLE POLE

PUSH BUTTON (SPRING RETURN)

CLOSING (MAKE)

OPENING (BREAK)

TWO CIRCUIT

(NO SPRING RETURN)

FLOW SWITCHES

CLOSE ON FLOW INCREASE

OPEN ON FLOW INCREASE

CLOSE ON RISING LEVEL

OPEN ON RISING LEVEL

PRESSURE OR VACUUM

CLOSE ON RISING PRESSURE

OPEN ON RISING PRESSURE

TEMPERATURE

CLOSE ON RISING TEMPERATURE

OPEN ON RISING TEMPERATURE

HUMIDITY

CLOSE ON RISING HUMIDITY

OPEN ON RISING HUMIDITY

THERMAL ELEMENT

ACTUATING DEVICE

THERMAL CUTOUT

THERMAL RELAY

THERMAL RELAY

THERMOCOUPLE

TRANSFORMER

VISUAL SIGNALING DEVICE

PILOT
(TYPICAL)
* ADD LETTER DESIGNATION

Appendix B: Common Notation

Symbol	Multiplier	Example
μ	Means: $^1/_{1,000,000}$ (Shift decimal to the left 6 times or divide by 1,000,000)	15.0μA = 0.000015A
m	Means: $^1/_{1,000}$ (Shift decimal to the left 3 times or divide by 1,000)	3.2mA = 0.0032A
k	Means: 1,000 (Shift decimal to the right 3 times or multiply by 1,000)	1.85kΩ = 1,850Ω
M	Means: 1,000,000 (Shift decimal to the right 6 times or multiply by 1,000,000)	34.5MΩ = 34,500,000Ω

Please refer to Appendix B, page 569, in your text for more common HVACR formulas.